Light as Experience and Imagination from Paleolithic to Roman Times

The tidal pull of light can shape an entire life.

—Ellen Meloy

Light as Experience and Imagination from Paleolithic to Roman Times

David S. Herrstrom

FAIRLEIGH DICKINSON UNIVERSITY PRESS
Madison • Teaneck

Published by Fairleigh Dickinson University Press
Copublished by The Rowman & Littlefield Publishing Group, Inc.
4501 Forbes Boulevard, Suite 200, Lanham, Maryland 20706
www.rowman.com

Unit A, Whitacre Mews, 26-34 Stannary Street, London SE11 4AB

British Library Cataloguing in Publication Information Available

Library of Congress Cataloging-in-Publication Data Available

ISBN 978-1-68393-094-5 (cloth : alk. paper)
ISBN 978-1-68393-095-2 (electronic)

∞™ The paper used in this publication meets the minimum requirements of American National Standard for Information Sciences Permanence of Paper for Printed Library Materials, ANSI/NISO Z39.48-1992.

Printed in the United States of America

for
Constance Joy
who is

Open my eyes
To the eyes of my wife upon me.
I am the tree at dawn. She is the light.

Contents

Acknowledgments

I am most grateful to David Massey who fanned the sparks of this book decades ago. And more recently, my friend's rigorous critique of successive drafts has not only been wonderfully sustaining but has improved this book. Without his casting a cold eye on writing too hot off the pixilated screen or his being unafraid to slash and burn in the interest of brevity and lucidity, *Light as Experience and Imagination* would have been half as alive and twice as long, to everyone's regret.

What a pleasure to thank Andrew Scrimgeour, a friend with whom I've traded writings since we were young. His warm but clear-eyed encouragement has sounded throughout my life. And his unwavering belief in my work has sustained me always. Without his faith in the significance of and untiring commitment to this project of light's human history, seeing me through challenges and in the end enabling me to publish this book, it would not be now in your hands.

Unending gratitude to my wife Constance Harmon Herrstrom who has been chapter by chapter a constant support and tough reader. Also, I would like to thank my friend Rebecca Mebert whose encouragement over a lifetime has been unflagging and whose notes-of-a-reader-reading have provided a useful perspective. Colleagues Alan Mallach and James Sherry have been most helpful at a critical juncture. While among those who have made this book possible, I owe a tremendous debt to Kenneth Stunkel, whose astute comments and wise suggestions as well as humanist spirit have made all the difference. My gratitude exceeds all words.

Introduction

> Light is too simple to be described, too glorious to be embellished, too mysterious to be explained, and, alas, too common to be prized.
>
> –The Reverend Henry Hunter, 1793

If I were to tell you about an encounter with an invisible force, an experience of a reality that has no substance but is visible and material, accompanies me everywhere instantaneously, fills me with joy and even ecstasy on occasion, possesses physical and psychological healing powers, and whose circumference is nowhere and center everywhere, you would assume, I suspect, that I'm talking about an angel or even God.

But no, this is the ordinary yet beautiful, mysterious light of a typical sunny day. And settled among the trees on a spring morning, it often has the presence of a god, demanding my affection and offering revelation. Growing up in California, I've long known this seductive presence. So this book tells the story of light seducing individuals down through the ages. It is a book borne from such experience, my breath taken, as I look out to a lone live oak in a dry field, by the light's mineral sharpness. Or my artist mother suddenly stopping me in her favorite redwood grove as if a cougar threatened, only to say, "look at that dapple-fall of light through the upper story."

Paradoxically, we do not actually see light. We see the glitter, glow, and shine that result from light interacting with things in the world. Our everyday experience is saturated with this natural light. We succumb to it and walk into the meaning that it makes while giving interpretive, imaginative meaning to light.[1] It is a force not only within our lives but within our culture. We will listen to individual testimonies of light-experience over time and explore each relationship revealed. By turns disturbing and affirming, these resonate down through the ages. Tracing why light arises in our lives and how it has affected us throughout our history, I follow the ways of our appetite for light.

We know the common and simple light pervading our daily life like the morning light settled in the woods. And we know the glorious and mysterious light occasionally brightening from all directions at once. Strangely, in the presence of such power, I have the sensation of belonging in some way to the light and the light to me. And I am convinced that this uncanny connection is not illusory. It is an optical tide as tangible as the trees, their leaves ingesting light, whom I've joined in some ancient reunion of light-grazers.

From our earliest history we have experienced light's force in our lives and in the life of the planet. We often experience an intimacy with its palpable presence to such an extent, in fact, that we've come to associate it with consciousness itself. We are "enlightened." Issues are brought to "light." A recognition grounded in our physiology. For our eyes have been over the course of some three billion years created by light. Having co-evolved with light, it pursues us even as we pursue it.

Figure 0.1. Light in Woods. ***David Baker, photographer (photo downloaded from web 11/24/09), but not the David Baker from Southhampton, Hampshire, U.K. (represented by Art Limited).***

We are so enamored of light that we collect specimens of it, savor and share them. It is this experience of a ubiquitous and personal light possessing the ability to effect wonder in us that is my subject. In *Light as Experience and Imagination* I tell the stories of how we've experienced the character of natural light throughout history, confronting and reflecting on it, asking not What is light? but How does light affect us? What does it do to us? What does it mean to us?[2] The answers to these questions expressed in many forms, from writing to painting, reveal the different valuations at different times that we have accorded light in our relationship with its stubbornly independent and utterly unique character.

And along the way we come to know light intimately, even as its remains elusive, changing through hours and seasons, forever improvising. It is both strange, radically different from any other phenomenon, and yet familiar. At times an alien and at others an intimate, I treat light as a character, even personifying it. Such a metaphoric strategy, I know, can be unsettling. However, this captures the dialog we have historically had with light and continue to have in our experience.

By "light," to be more precise, I simply mean our common experience, such as sunlight sifting through leaves or glancing off snow, revealing and connecting us to the world of objects—natural or visible light. The light seducing us by inducing various moods, and connecting us to each other as well as to the eternal. This natural light is a mix of direct and indirect sunlight. The latter is most common in our experience and consists of skylight, which is light scattered from the air and clouds as well as reflected from the ground. We receive little direct sunlight because our atmosphere with its molecules of air and water absorbs much of a sunbeam's light. It also scatters light, redirecting it out of a beam.[3] The natural light of the sun is largely brightness with variations of intensity, which is what the basic mechanism of our eyes responds to—luminance. The sparkle and glitter of white light being a fundamental human experience, it penetrates the deepest recesses of our soul.[4]

No one is immune to this familiar and strange being in our world, as an ambitious boy born in Scotland in 1741 was to discover. After he completed his education and took ordination, the Reverend Henry Hunter accepted a call from a congregation in London, having no inkling of what was to befall him. Routinely working on his sermon one morning, he was visited by light. Preached in London on November 14, 1793, Reverend Hunter's sermon confronts this light and sums up its accessible yet paradoxical character as "too simple to be described, too glorious to be embellished, too mysterious to be explained, and, alas, too common to be prized."[5] An epitome of our responses to light, I'll use this as a map to its experience throughout my book. In hearing individual testimonies, we'll discover light's generous, independent, and disruptive sides.

At the same time, we'll preserve the reverend's awe and wonder in the presence of light, revealed in his admission that light in the end will always be a mystery beyond description and explanation. While light is a common universal phenomenon to which we owe our very existence, as does all life except the sulfur-eating bacteria in the blackest depths of the ocean, and while light's glorious beauty is universally acknowledged, it remains a mystery. Just when we arrive at a description, our words dissolve in wonder. I can identify light's generosity, filling all the nooks and crannies of our world, its independence of our earthly home as well as of us, and its power to deny or sustain life, but these traits do not exhaust the character of light.

So I write about ordinary light. It's often glorious and sometimes mysterious character constitutes our very lives. Acting as receptors, our eyes are helpless. They receive, then absorb and process. Light pummels and fills them. This becomes our "lightscape," the familiar world that we land on when we emerge from the womb, our everyday human space and time created by light.[6] The luminous skies of the Mediterranean place us in a different space, for example, than the muted skies of the north Atlantic. As we move through the world, light provides continuity, one area of our ever-changing lightscape flowing into another. Although not a child of time itself (since time stops at light-speed), then, light creates personal time. In fact, light is nothing less than the author of our world, creating all and participating in all, indifferent and immanent.

The perceived moods and movements of this character will be our subject in the chapters to follow. I'll explore our intimate relations with light and its "social life" as it affects our senses and illuminates the people, places, and objects of our world. Moody and changeable, light affects us profoundly. Light deprivation can make us sick. On the other hand, excess of light can be torture. And between these extremes, which is the arena of our exploration, we grant value to various aspects of light's character, whether radiance filling us with hope or shine that seems a "spiritual" presence.

In exploring our experience of visible light, I take inspiration from the eighteenth-century tradition of "natural history." This is a history written by amateur "professionals."[7] And I take the naturalist's approach in exploring the human experience of light. It is reported encounters with light that are the objects of this natural history. These eyewitness testimonies of natural light down through history constitute my "specimens." They have been left by individuals who value their experiences of light enough to want to tell us about them. We'll see how, in fulfilling this desire, they create a personal portrait of light, as they use various means of expression, such as painting and poetry, architecture, and prose to communicate their experience.

Encounters with light can be exhilarating or discomfiting. A certain slant of light can trigger joyous memories or frightening ones. In either case, the experience of light is as intimate as a dream. "All the forms of light quiver in my essence," the modern French poet Paul Valéry reported.[8] Paying particular attention to how eyewitnesses like the poet prized and described, as well as embellished and explained, the unique and ubiquitous light of our world, revealed in their speaking about light, mine is the story of their light-experiences.

And beyond the personal, "each age brings with it its own light," as Henri Matisse observed.[9] Not just for the painter but for all. Each witness creates their own specific portrait of light within what we could call the human light-climate, each testimony of light-experience as valuable as that of a previous or succeeding period. Change from age to age means a change in the relation to light, expressed in the sum of portraits made by individuals in that age. As an individual's fascination with light finds its expression, their observation and interpretation, as well as their cultural memory and mythology, mix. From ancient to modern light, the character of light changes as a function of the experience of light within a culture, each age creating a legend of light that captures the imagination of a whole society.

From earliest times, for example, these legends often made light a conduit for power. Light was imagined as a manifestation of a transcendent realm or the gods themselves. Consequently, a major shift in value occurred when light was emptied of its divinity and became an inanimate object.[10] This occurred at various times in the experience of individuals but was not culturally consolidated until the eighteenth century when the Western mindset took a turn from the religious to the secular imagination.

To see how the figure of light changed over time, we read carefully the languages that each age devised to describe this ever-elusive character. Sometimes light mocks our attempts to prize and describe it, yet always encourages wild embellishments and abstruse explanations. We have explicit accounts of encounters with light in writings, such as poetic or philosophical meditations. And we discover implicit ones in buildings and mosaics. Like me, their authors want to "read and repeat" the "eloquences of light's faculties."[11] My history relies on these accounts, which become the "languages" of light. And of each I ask: What is, as I call it, the "core" experience? Why did this arise in the life of an individual, and how did it affect the life of the individual?

To answer these questions entails exposing the individual's fundamental encounter with light that commonly begins as a single image. This image may be the special radiance of dawn or blaze of noonday sun. It can be mundane, such as a flash of lightning in the distance, or an epiphany so pure and unlike any ordinary light that it possesses the force of some overwhelming spiritual power.

Either way I pay close attention not only to the personal and historical circumstances that birthed a specific encounter with light, but to the metaphoric interpretation that an individual gives this incident. Encounters so interpreted can resolve personal contradictions inherited from the culture, forging a new identity, or create conflicts with a cultural tradition. My conclusions result from an exploration of the force and consequences of a person's "core" experience, such as how it haunts the memory, becomes an obsessive symbol, or gets elaborated in a personal myth. These can take the form of philosophy or religion, science or poetry, and have a profound influence on others, such as Homer's light inseparable from the heroic ideal.[12] For it shapes the culture, which in turn shapes subsequent personal experience.

Not only does the experience of light affect individuals in how they understand themselves but in how they understand the world and the universe itself. The third-century Roman philosopher Plotinus, for example, extrapolated from his core experience to light's own origins and life story an entire vision of how the cosmos operates, linking the mundane to the supramundane and vice versa. How such stories are linked in a given age and over the ages becomes the natural history of the human experience and imagination of light. Their resemblances and their differences tell us about changes in the character of light that is its history.

Common sense tells us that we discover light as we sit in our backyard or move about in the world, that it is something found just as we would find a stone, a purely natural phenomenon. And we merely record its effect. In reality, however, given our ongoing dialog, we "construct" light. It is, in fact, a complex cultural document embodied in artifacts such as poems and paintings that require interpretation. Just as my eye constructs an object by making a map of it, a prior act of imaginative construction takes place when I encounter any light. My visual apparatus selects certain photons and not others, and my psychological and cultural inheritance predisposes me to select certain qualities and not others.

It is these acts of construction, rooted in our physiology and extended by our culture, that we call our experiences of light. And as I've suggested, we have developed or adapted various languages to express these that lie between two extremes. On one end is painting, flexible, capable of expressing a wide range of experience. Invented perhaps about the same time as speech, painting was old by the time of the Stone Age hunters we'll meet in our first chapter, who drew and painted animals on cave walls and the ceilings more than thirty thousand years ago in what are now France and Spain. On the other end is the language of Euclid's optics, relatively inflexible. But our experiences find expression in a range of languages between these poles, from prehistoric cave paintings and Egyptian poems of 2500 BCE to the dialogs of Plato.[13]

We satisfy certain needs, then, by re-imagining our encounter with light. While we live in the light, our constructed light lives within us. It can be intimate, even sexual, a manifestation of another world that comforts or challenges us. And this experience, as we explore its meaning for our lives, turning to metaphor and symbol, can either be sanctioned by tradition or subversive of tradition. Old religions have been bolstered in their authority, for example, or shattered. Philosophies have been extended or upended. In any case metaphors of light carry force, whether light is re-imagined as spiritual, political, or intellectual power.

Reassuring or revolutionary, all witnesses to light's character, from painters to scientists, employ the fundamental language of image, metaphor, and symbol for personal or social ends. And my project is to interpret this "rhetoric of light." Painting, science, and philosophy, all present metaphors that rely not only upon discovery and invention, but on the discoverer's systematically recreating the world in the imagination. Resisting cultural pressures to divide the human imagination, light stands at the crossroads of art and science. Its unique character drives the scientific as well as artistic and religious traditions that make us recognizably human.

On the journey from the prehistoric to classical ages, we'll see how individual encounters with light trigger discovery. From prehistoric shamans in my first chapter to Roman artists and writers in my last, light displays all facets of its character, while at the same time asking us questions and revealing knowledge. Examining specific testimonies about aspects of light's character and its action in the lives of individuals, we find ourselves in awe of this interrogator and revelator.

This book focuses like a burning glass on our encounters with light and ways of making sense of these, beginning with the earliest records of our experience of light—Paleolithic cave paintings and mythic tales. I am concerned with the personal and cultural impact of light, then, not with tracing the development of optical science as it strives to discover its narrowly defined "essential" nature. As age follows age, our relationships with light, wave on wave of testimonies to light-experience, do not get closer to any single truth.[14] But rather, we glimpse more facets of light's character and inherit a greater understanding of its richness.

The chapters that follow highlight the singular individual experience, exploring our conflict with, resistance to, and acceptance of light. Because it is like no other phenomenon in our universe—familiar yet profoundly strange, revealing itself as it hides itself—light is new for every generation. In these chapters, we'll meet individuals who give us rich reports of their personal encounters with light, revealing many traits of its common and simple, glorious and mysterious character. And just when we think we truly know light, we'll find ourselves glimpsing previously unknown facets of its character.

NOTES

1. Light masquerades as a penetrating beam or, Hans Blumenberg exclaims, "a guiding beacon in the dark, an advancing dethronement of darkness, but also a dazzling super-abundance, as well as an indefinite, omnipresent brightness containing all: the 'letting-appear' that does not itself appear, the inaccessible accessibility of things. . . . Light is intrusive; in its abundance, it creates the overwhelming, conspicuous clarity with which the true 'comes forth.' . . . Light remains what it is while letting the infinite participate in it; it is consumption without loss. Light produces space, distance, orientation, calm contemplation; it is the gift that makes no demands, the illumination capable of conquering without force" (Blumenberg, "Light as a Metaphor for Truth," 31). "All types of light experiences," as Mircea Eliade says, have in common that they "bring a man out of his profane universe of historical situation, and project him into a universe different in quality, an entirely different world," often "transcendent and holy." But the meaning of this light, whether secular or holy, is, "on the one hand, ultimately a personal discovery; and, on the other hand, each man discovers what he was spiritually and culturally prepared to discover," but either way is "produces a break in the subject's existence," often realigning the "world of the spirit, of holiness and of freedom" (Eliade, "Spirit, Light, and Seed," 2; and see in this article his "Morphology of Photisms" found world-wide in experiences of "inner light," 23–25).

2. Questions central to the anthropology of light, a discipline which adds another, "How do we use it?" (see Bille and Sørensen, "An Anthropology of Luminosity").

3. The light we see and physically respond to is a narrow band (about 1 percent) of the broad spectrum of "light" that ranges from low-frequency infrared to high-frequency ultraviolet rays (electromagnetic radiation). All its visible frequencies vibrating together like the sun's, to which our eyes are tuned, are known as "white light."I leave it to others to explore light's separate notes within this band—colors (frequency), an immense subject in itself, and polarization (angle), which unlike bees we are blind to without the aid of polarized sunglasses. Though colors carry luminance information, it doesn't matter for our purposes because the part of our brain that analyzes luminance is colorblind (Livingstone, 38). For a history of color, see John Gage, *Colour and Culture* (London: Thames and Hudson, 1993); of color in painting Faber Birren, *History of Color in Painting* (New York: Reinhold, 1965), and of pigments, Philip Ball, *Bright Earth* (New York: Farrar, Strauss and Giroux, 2001); for an influential treatise, see Roland Rood, *Color and Light in Painting* (New York: Columbia Univ. Press, 1941), and for a useful bibliography, David Batchelor, *The Luminous and the Grey* (London: Reaktion Books, 2014), 105–07.

4. The proof that we have always had an affair with light is not merely that we have always named light itself, but that we have named specific aspects of our experience of light. Anglo-Saxon *leoht* is allied with *leoma*, a gleam. This word, *leoma*, witness to an ancient preoccupation with one aspect of light's character, its luminance or "gleam," is a remnant as intriguing as the fossils of ancient clams that Leonard da Vinci found three hundred miles from the sea at the top of a mountain. This word's history extends back to its Indo-European root **leuk-*, which tellingly carries into the Greek, Romance, and Germanic language families (*Dictionary of Untranslatables*, 577). Our modern tongue, continuing the tradition, blossoms with words that strive to name just this aspect: shine forth, gloze, glint, glance, flash and fulgurate; bedazzle, glimmer, spangle or coruscate; glister, bicker and luminesce; lustrous, orient, aglow, fulgid, effulgent, garish; shiny, flashy; lambent and scintillant.

5. *Dictionary of National Biography*, X, 287; "too simple . . . common to be prized," Hunter, *Sermons*, 2: 279.

6. The womb itself is not totally dark, however, because some light penetrates our skin. We're acutely aware that our lives are governed by the rhythm of light and dark, which is true at the most basic level because our pineal gland, acting on light signals, regulates our sleeping and reproducing. We're not the only creatures whose clocks are governed by light of course; a species of mosquito "reads" light, and a butterfly exists with light sensors in its genitals. For studies of the physical effect of the sun on our lives and the world, see Howard Harold Seliger, *Light: Physical and Biological Action* (New York: Academic Press, 1965); Peter A. Ensminger, *Life Under the Sun* (New Haven: Yale University Press, 2001); and Peter Hegemann and Hartmann Harz, "How Microalgae See the Light" (*Microbial Responses to Light and Time.* Eds., M.X. Caddick, et al. Cambridge: Cambridge University Press, 1998).

7. For example, the clergyman Gilbert White's (1720–1793) *The Natural History of Selborne* (1789). Though not a professional, he was a pioneering naturalist with a claim to being England's first ecologist (Gifford, 10). On the meaning of "natural history" and on White and Thoreau as natural historians, see Gifford's "Introduction."

8. "All the forms of light quiver in my essence!" quoted by Plummer, *Poetics of Light*, 137, 193. And as the philosopher Gaston Bachelard observes, "the painter experiences the revelation of the world through light too intimately not to participate with his whole being in this ceaselessly reiterated birth of a universe" ("The Painter Solicited," 29).

9. "Each age brings with it its own light," quoted by Jack Flam, *Matisse on Art*, 156.

10. This progression from ancient to modern may suggest that light's portrait changed in the imagination from a spiritual to a physical being. Northrop Frye, for example, concludes that the evolution of the historical experience of light passed through three phases: the first begins with God, an "articulated" light; the second with man, an inner light; and the third with nature, a physical light ("The Light of Myth," 8). But such a generalization does not do justice to the contradictory portraits of light within the same light-climate or to the lively variations of personal experience. More nuanced is Arthur Zajonc, who sees an arc from what begins as a "lively, soul-spiritual" light that then divides into optics and psychology. We move in this scheme from the ancient "psychological space" to the later "mathematical imagination of light" (*Catching the Light*, 24, 114, 270), and return to the modern "spiritual conception" of light in philosophy (e.g., Rudolf Steiner), quantum physics (Max Planck), and art (e.g., Robert Delaunay). Clearly, along the way it was a disruptive "energy and force" when, according to the architect Henry Plummer, "light itself was experienced in a new way" (*Masters of Light*, 20).

11. Wallace Stevens, "The Pure Good of Theory," IV (*Collected Poems*, 333).

12. I use "image" and "core experience" to designate what Martha Blassnigg defines as an "action set out by, or in, (human) consciousness"; the metaphoric/symbolic transformation of the image is a "quality encountered in both visual perception and cognitive activity of imagination as a co-creative action." Thus, like her, I want to focus "on the particularities of the quality of light in its interactions with imagination." In this she follows Bergson, understanding the "image" as an "indicator of the full range of sensory and imaginative experience involved in image perception and creation," situated between the "realist notion of a 'thing' and the idealist notion of a 'representation.'" It recognizes, that is, the intrinsic entanglement of matter and mind, inside and outside, virtual and actual (*Light Image Imagination*, 12, 21).

13. These languages, listed roughly in their order of origin below, overlap and interact in time as each develops, such as architecture being shaped by religion and Western painting by optics. They also intermingle within an individual witness like the Roman poet-philosopher Lucretius in whose works myth, history, philosophy, and science cannot be disentangled. Any one language merges into the other:

1. Painting/architecture/sculpture
2. Poetry/mythology-religion/philosophy
3. Optics/photography/film (modern world)

These three clusters, the visual (or plastic), the verbal (or conceptual), and the formally coded, whether by mathematical diagrams or by light-sensitive media, enable experiences of light to be expressed. Each of these languages is shaped by the nature of its medium, whether words or plastic art, such as pigment or limestone, or composites, such as treatise (and later film). Accordingly, they differ in the aspects of light's character they best capture. Words differ greatly in their expressive ability, of course, from the surfaces of pigments or buildings. Whereas poets manipulate metaphors and storytellers events, artists convey their experiences by altering the way surfaces reflect and scatter light.

Using the materials of their respective arts, that is, artists discovered ways of conveying their perception of light early on by "alterations" of (1) surface *layout* (i.e., reliefs, sculptures, architecture) and (2) "surface *reflectance*" (e.g., painting). Later, beyond ancient times, they discovered the possibility of (3) alterations of "surface *illumination*" (e.g., theatre and spectacle, photography) and (4) alterations of what we might call "*refractance*" (e.g., film) (Gibson,

224). These modulations only evoke the experience of light, which does not reside in the material itself, of course, but in our response. Pigments, that is, are metaphors.

Painters were not alone in developing a rhetoric of light, ways of persuading us to accept the power of their experience and passing their codes on in the teaching of traditional techniques as did the Greek painters. Euclid's rays, for example, were shorthand borrowed from geometry for an experience we all share. Each of us has seen shafts of light from clerestory windows slanting through suspended dust particles, such as in a cavernous train station like Chicago's, but Euclid took the next step and generalized these observations. In doing so he captured something of light's essential formal character, its habit of traveling straight paths, and invented a code that was easily manipulated for further discovery (Bronowski, 46).

14. David Park's *The Fire Within the Eye*, Richard Weiss' *A Brief History of Light and Those That Lit the Way*, and Bruce Watson's *Light: A Radiant History from Creation to the Quantum Age* all tell the story mainly, unlike mine, of progress in optical science with asides on personal experience. Each of these is more concerned with the question, "What is light?" than with "What is our experience of light?" Nonetheless, I am particularly indebted to Park's absorbing history of the investigation of light's behavior, as I am also to Arthur Zajonc's exhilarating study of the "entwined history of light and mind," *Catching the Light.* I've learned much from these two scholars. Like mine Park starts with experience, but his book is principally a history of "thought about light." And like Zajonc's mine is a biography of light, but tells a different and complementary life story.

Chapter One

Hunter Light of the Sky

The Paleolithic Era

Just as they were with the animals, our earliest ancestors were compelled to deal with light, all inhabiting the same lands. So how did they come to terms with the "common" and "simple" physical fact of light? To answer this, we go back to the beginning as related in a tale from prehistoric times, which introduces Raven, a bird-man who steals the light. This story emanates from a shamanistic society of hunters and gatherers in the early Stone Age era known as the Paleolithic (c. 35000–10000 BP [Before the Present]).[1] A tale expressing these peoples' experience and imagination of light and Raven with his "shining eyes" and ability to transform himself stands in for the shaman, a central figure in most prehistoric hunter-gatherer cultures. At home in the sky, Raven is the bringer of light, as the shaman tells his tale, testifying to his experience of a material light existing independently in the world.

In the beginning was Raven, so begins a version of the story told by the Haida, a tribe of hunters and gatherers who make their home on the coast of British Columbia.[2] And in the beginning was "all consuming dark" because an old man who lived with his daughter, kept "all the light of the universe" in the smallest box nested within an "infinite number of boxes." But Raven didn't like this world of darkness, so he decided to steal the light. Contriving to be born of the old man's daughter, he tricks her into giving birth to a strange-looking baby boy with a few feathers and shining eyes. Once born, Raven wins her father's affection. And the old man finally gives into his begging to see what is in the boxes and "hold the light for just a moment."

As they near the last box, a "strange radiance, never before seen, began to infuse the darkness of the house." When the old man reaches in and lifts out the light, Raven immediately snaps it up in his beak and swoops out into the "huge darkness of the world." And in that moment the "world was at once transformed." Raven sees "broken reflections" off a river for the first time and, letting go a last "piece of light," sees it float and start "up over the mountains lying to the east," its "first rays" catching a house by the river.[3]

Most notable in this myth, light is a material substance that can be stolen and broken into pieces. Also, light and by implication darkness have their own physical existence. They are independent of other objects in the world and of any associations with heat or cold. In the beginning each exists in itself. The tale also reflects an experience of light as bound up with darkness, just as life is bound up with death. Furthermore, neither is subject to cause and effect, the sun rising or setting. Light precedes the sun (as it does in Genesis). And darkness precedes the sun's daily disappearance. Both light and darkness, that is, retain their individual status, though like twin brothers they are complementary.

So this mythic tale gives us a hint about how the character of light was experienced by prehistoric peoples. We learn from it that inasmuch as light is granted its own existence, the "common" light of their everyday experience is prized. It is a physical entity to be coveted and, clearly, not taken for granted as we do, which is the mark of "In the beginning," where past origins give meaning to present experience. At the same time, light is a "simple" entity, appearing in the world just as familiar plants and animals and stones do. And it is this character existing independently apart from any source, the sun or moon, that begs to be described. Prehistoric peoples tell us about their encounter with this light not only in myth but in the language of prehistoric art. Namely, the Paleolithic cave paintings of France and Spain we'll visit in this chapter, and the Neolithic megaliths of Britain and Ireland in the next.

These painters and builders document an experience of light not only as substance, as we've seen, but as a body. And we'll see this light-body emerge in Paleolithic cave paintings. Light cannot be separated from this physical or "carnal" experience of it, which constitutes our deepest experience as humans, a fundamentally unconscious one.[4] Thus prehistoric cultures do not embellish light as a "glorious" phenomenon or explain it in tales as a "mysterious" one, consciously abstracting its aesthetic impact or symbolic meaning, which we'll see in later chapters. Instead, they welcome its body into theirs through vision induced by hallucination, as does the Paleolithic shaman; or by ushering it into theirs through eyes wide open, as does the Neolithic *seer* of our next chapter.

CRO-MAGNON ANCESTORS, THE IMAGE-MAKERS

So we discover that our earliest ancestors had to come to terms with the physical light that aided and endangered, enabling them to hunt prey but also prey to hunt them. This common and simple character demanded of them accommodation. Our investigation of their experience, then, takes us back to the beginnings of what we call "modern" man. Our prehistoric ancestors of the Paleolithic and Neolithic eras, people known as Cro-Magnons, migrated to what is now western Europe some forty-five thousand years ago.[5] They are us (*Homo Sapiens*), the same look, the same intelligence. And even the earliest of these peoples, Paleolithic hunter-gatherers living in shamanistic societies before the invention of agriculture, supported specialists, painters, and storytellers, such as the shaman who was often skilled at both. Furthermore, these Cro-Magnons, having migrated to lands already occupied by Neanderthal peoples for some thirty thousand years, invented elaborate rituals, which possibly provided the ceremonial context for their cave art.

The period during which our forebears prevailed over the Neanderthal from roughly 45000 BP to 35000 BP has come to be called the "Paleolithic Transition" or "Revolution" centered mainly in what is now southwestern France and northern Spain. During this transition, Cro-Magnons underwent a stressful psychological and technological revolution that gave rise to a Paleolithic "Creative Explosion." With their displacement of the Neanderthals came a new kind of consciousness. This was not a matter of superior intelligence but of a particular kind of consciousness. A more complex neurological structure enabled these people to introduce elaborate burial practices and image-making.[6]

This means that Cro-Magnons could entertain mental images independently, projecting them outside themselves and manipulating them. Furthermore, these could be extended in drawing and painting to a two-dimensional representation of a three-dimensional object or being. Having a fully modern consciousness in contrast to the Neanderthals, these men and women had the ability not only to contemplate mental images but produce them. It is this pivotal difference between the two species that triggered the "efflorescence of image-making," a veritable explosion of cave painting. Painting never surpassed, as Picasso once acknowledged, for sheer artistry and emotional impact.[7]

Because Cro-Magnons had the ability to make and manipulate images, the Paleolithic painters and, as we'll see in the next chapter, the later Neolithic builders were equipped to communicate their experience of light. The shaman expresses an intuition of the bodily fullness of light within, which mirrors the sky without, in the language of paint (as well as in engravings and highly polished sculptures), whereas the Neolithic culture-hero, a "seer" whom we'll meet later, expresses his perception of light, which resonates with the sun, in the language of architecture. What we learn from these respective languages marks the beginning of our history of the experience and imagination of light simply because we have no evidence of man's encounter with light prior to this age. And prehistoric Cro-Magnons describe an experience of light before it is divorced from the sensing body and abstracted in later ages, which enabled its aesthetic perception as "glorious" and symbolic investment as "mysterious."[8]

Cave painting, then, constitutes the earliest record we have of the human experience of light. And what we find recorded here is the most fundamental aspect of light's character, its sensuous material body existing independently of the world and any source. Painting was central in Paleolithic hunter-gatherer societies of western Europe, and the shaman as a specialist in altered states and very likely as painter was at its center. His ordinary consciousness apprehended the experience of daylight as a palpable encounter. But his extraordinary consciousness given to visions consolidated this character of light as a bodily presence that he made concrete on cave walls.[9] By means of these visions the Paleolithic shaman retrieves the sensuous light of the world. It is to the light-filled sky of his ordinary, external world, as we'll see, that he must journey in order to accomplish his task. So he becomes a bird, as in Raven's tale, an animal who mediates between earth and sky. But this journey is also one to the internal world of vision. What he finds outside and inside himself is a material light that he can touch like his own body.

THE SHAMAN'S TRANSFORMATION AND THE LIGHT-BODY

Shamanism first appeared in the late Paleolithic era with our new consciousness. And given shamanism's "antiquity and ubiquity," as David Lewis-Williams argues, it is very likely that the Paleolithic hunter-gatherers of western Europe practiced it in one form or another. The shaman was a religious specialist, a professional who exploited altered states of consciousness often induced by hallucinogens or sense deprivation, such as being enveloped by the utter darkness of a cave for an extended period of time.[10] In his altered state the shaman undergoes a visionary journey, encountering an animal guide who aids him in achieving a light-body. By this the shaman gains power, enabling him on his return to his community, to serve as a healer.

Our shaman of the cave paintings ascends like Raven into the other world of light, which is and is not the sky, becoming the body of light with the aid of his animal guide. Subsequently, he descends into the world of darkness in the cave to consolidate and record his vision. To complete his journey, he returns to the ordinary world of his tribe as a light-master. To prepare for its first stage, the ascent, he transformed himself into a bird-man. Astonishingly, we have a record some seventeen thousand years old of this transformation. On a cave wall in southern France in the Dordogne, discovered in 1940 by a dog named Robot and four boys, a Paleolithic shaman drew an uncanny figure. It is a man with a bird's head and bird-like hands falling backwards as if in a trance (see Figure 1.1). He appears to be a bird above the waist and human below, while the bird-head on the staff standing beside him echoes his own head.

Clearly, this image is special. It is the only human-animal figure out of almost two thousand representations in Lascaux (c. 17000 BP). And the fact that the picture is placed at the bottom of a natural shaft below the main floor of the cave in a place that is difficult and awkward to get to also points to its significance. Just to position himself for drawing this image involved a physically harrowing journey. The shaman had to lower himself down a shaft with a rope in the dark and then rekindle his lamp in order to draw this bird-man. We can't help but suspect that this niche was reserved for the shaman alone, his sanctuary set apart from ordinary men. Paintings in a shaft difficult to access were evidently not public. And we see here the transformation of a

Figure 1.1. Birdman of Lascaux Cave. c. 20,000 BP. ***From International Business Times (12/2/16), "Lascaux cave paintings still hide secrets more that 70 years after their discovery," Léa Surugue. Accessed 3/10/17. http://www.ibtimes.co.uk/lascaux-cave-paintings-still-hide-secrets-more-50-years-after-their-discovery-1594447***

shaman into his "spirit helper at the moment of death," a bird-man like Raven. A special "death," it is transformative for the shaman not terminal, as it is for the eviscerated bison depicted beside him. It means traveling to the "spirit world" via an altered state of consciousness, a passing through the "portal to shamanistic status" and becoming the light-body.[11]

Whatever the technique of achieving an altered state, the mind is "inclined to invent light." For the shaman this is an internal light that chimes with the external light of ordinary consciousness. We see, for example, "concentric circles or endless chains of brilliant dots," known as "entopic" motifs, deriving spontaneously from the physiology of the eye at the outset of an intensified state of consciousness. Similar motifs occur in the Apse of Lascaux, which has special significance as the "most sacred place" in the cave. The San of South Africa, who share with the cave painters a shamanistic culture and whose art reaches back into Paleolithic times, use these motifs as well in their rock engravings and paintings. Lines called "threads of light," usually a red line accompanied by white dots appearing and disappearing in and out of the rock, represent the supernatural power that they harnessed in order to go on "out-of-body journeys." At their outset, the shaman begins "to climb the thread" into the sky. Associated with a beam of light and falling star, these threads are incandescent filaments of light.[12] Though the body generates such images, it is the interplay of individual and culture that imbues them with meaning.

In the final stage of the shaman's intensified state of consciousness, at the end of his journey, where his natural and supernatural bodies unite, we meet the character of light as it manifests itself to the earliest Cro-Magnon in its purest form. And it is body. As the shaman takes on the body of the animal and becomes the animal, so he takes on the body of light and becomes the light. This "carnal" light experienced by Paleolithic people is simultaneously celestial and chthonic. The shaman lives in the sky and the earth. This brings into conscious view the fundamental character of light, which as humans we experience below our consciousness. Ascending as a bird and descending into earth as a bear, which we'll meet, the shaman experiences "light coming into being."[13] This in turn is bound up with his experience of the animal, as it comes into being on the walls of his cave by means of paint and light, inseparable from his experience of himself as it comes into being in his vision.

But what has this light-body achieved in his visionary union with light? Among the Inuit, the shaman has an "illumination" or "lightning," which is called the *qaumaneq*. One Inuit told the great Danish Arctic explorer Knud Rasmussen that "every real shaman has to feel an illumination in his body, in the inside of his head or in his brain, something that gleams like fire, that gives him the power to see with closed eyes into the darkness." This "mysterious light which the shaman suddenly feels in his body," resulting from an altered state of consciousness, "enables him to see in the dark, both literally and metaphorically." Such is the carnal experience of light.[14]

Having become the light-body, light shines from the shaman's eyes. And this power enables him to see into the reality of things. The Australian master of the initiation is known by "the light that shines from his eyes," the aspirant shaman receiving this light by ingesting rock crystal believed to be "solidified light."[15] And the same was true no doubt of the Paleolithic shaman. He possesses these eyes in common with animals of power, depicting them on his cave walls, such as the single outsized eye of a bison that dominates one wall of the Black Salon at Niaux in southern France, which arrests us with its intense gaze. Outlined in bold black strokes on a pale body, this stunning eye is echoed by a bison eye at Lascaux. But here figure and ground, light and darkness switch, as the eye shines out from the saturated black of the animal's head.

This is the puissant eye that can, it seems, penetrate the cave darkness just as do the eyes of the shaman himself (a.k.a. "Sorcerer") gazing on us from the cave wall of Trois Frères, Ariège, France. By dramatically emphasizing the round pupils of this figure, outlining them with black circles, the artist brings us up short. We are confronted with the head of a shaman twisted full face toward us with "wide-staring eyes."[16] Here the shaman asserts the power of light emanating from the body, his body of light.

The shaman identifies with his painted animal counterparts. But these are not by and large as we might assume the ones he commonly hunted. Instead, they were those endowed with extraordinary power as mediators between the natural and supernatural worlds. These animals served not only to guide the shaman in his quest for the light-body but to aid him in finding the means to communicate this encounter in painting. Such are the formidable bears about the size of Kodiak grizzlies or larger that roamed his country and held a unique status in hunter-gatherer culture. Living in the caves before the artists, they left claw marks and skulls, so a special relationship was obtained. Bears were intermediaries between humans and animals. This persists in modern shamanistic cultures. And in Haida culture, which gave us the tale of Raven stealing the light, bears were believed to be men in animal form. Given the Paleolithic shaman's circumstances, this relation is no surprise. After all, they were two-legged (bears often rising up on two legs) like birds and the shamans themselves who over the centuries took turns with them occupying the caves.[17]

More important, perhaps, they were ancestral artists. For just as the bear engraved his mark on the cave wall by raking it with his claws, so the shaman engraved his "flutings" by dragging his fingers through a soft coat of clay on the wall in certain places like a child's finger painting. And the image of the bear is prominent in the oldest cave. Throughout the caves, shaman engravings often mimic the bear's rakings. Fittingly, he appears to honor the ur(sus)-artist bear by placing an adult bear's skull on a block of stone that had fallen to the floor in a chamber of Chauvet cave.[18] Possibly identifying with bears making their marks on cave walls, then, the shaman discovered the technique he employs to express his encounter with light. He seems to acknowledge this debt, in fact, by occasionally stenciling his handprint signature on a wall as if it were the bear's. And the rhythmic placement of some of these prints that trail a painted bear who tracks off along one wall suggests they are equivalent to the bear's tracks. A curved row of red dots like San "threads of light," which signified power to carry out a vision journey, accompanies these prints, further suggesting that the Shaman-painter has left his signature, just as the bear leaves his tracks behind him. By taking on the bear's powers, the shaman extends what the ancestral bear had begun—making meaningful marks on cave walls.

So animals of power not only aid the shaman in achieving his light-body vision and transformation, but they are possibly instrumental in introducing him to the language that he'll use to consolidate his vision—engraving and painting. As a result, animals are the primary subject of the Paleolithic painter's wall art, which distinguishes it from subsequent artistic traditions and gives it a certain strangeness. Animals are the shaman-painter's primary subject because they are his guides and teachers. Accordingly, they too possess light-bodies, such as the bear painted on a wall of the Chauvet cave (in The Panel of the Little Bear) alive with light owing to the artist's making his simple outline drawing around the white fossiliferous veins of a limestone wall.[19]

Enlisted from the natural world to guide him in the supernatural world of vision, Bear and Bison aid the shaman after he ascends to the sky. By the same token, having been given a vision and initiated by them into his art, they descend with him into the earth where the shaman gives them a visionary life on the walls of a cave. By this act he solidifies his own vision so that he can be its audience, fixing it in paint in order to meditate on it. This externalizing of his vision, its physical consolidation, allows him to draw on its power at will whenever he enters his sanctuary.

THE SHAMAN'S ART OF LIGHT—THE CAVE PAINTINGS

If the shaman's vision is largely an unconscious project, his painting is a highly conscious act. He may have been introduced to the idea of making marks on a wall by Bear, but over millennia the Paleolithic artist extended these marks beyond the bear's simple rakings and perfected his art. This was driven by necessity. Giving substance to his light-vision—his transformation and assumption of a light-body—in paint was essential for appropriating its power. To this end he developed and employed refined techniques and sophisticated pigments in order to convey the light-experience that charged his images with power. His fundamental experience of light's character is, namely, the bodily apprehension of light being inseparable from light itself. And in his cave sanctuary, in near-inaccessible niches that only he could visit, he draws on the power of these images. By them he brings healing to his tribe and the assurance of food, the return of commonly hunted prey.

The shaman's subject, of course, is ostensibly animals of power, his visionary guides and instructors such as Bear and Bison. To these he dedicates the techniques and materials of his art. Preoccupied with animals over millennia throughout all the caves, which stretch from southern France down into and across the north of Spain to the west, the breathtaking paintings strike us at first as naturalistic. Yet on a closer look they do not employ perspective, framing, or any consistent scale. In this sense, they approach the expressionistic. Clearly, the painter in representing the animals defines his relationship to them. But as with the bear of light and the bison of light, this serves to reveal his relation to light itself. He used his skills not only to consciously depict the play of light on a sleek hide and the turn of a bison's head from shadow into light, which resulted from hours of observing animals in every detail for his own survival, but unconsciously to express his deepest and most intense experience of light.

To fully understand this experience, we'll follow the Paleolithic shaman-painter down into the cave itself because it was not enough to have a vision. As we have seen, he was compelled to give it external reality. Clearly, many images were not painted while the shaman was caught up in a vision, but by means of them he realized his vision on the cave walls. These visionary images are as visceral for him as the cave walls themselves. Embodying his mental light, they make palpable the light-bodies of his animals. By means of them, caught in the act of painting, he consolidated his power within the tribe for whom the animals would seem to appear magically on the cave walls. So it is into this place of darkness, the womb or "entrails of the nether world," that at great personal risk he descends from the sky world in order to record his experience.[20]

The painter succeeds in projecting his images onto the cave walls by using techniques and materials that exploit the flickering light and undulating walls of the cave setting, his constant companion. He often places images strategically to take advantage of a varying light, for example, and uses natural rock formations to emphasize his animal forms. For he views the cave walls themselves not as passive vehicles for his images but as active partners with him, continually animated in the quavering light of his torch or oil lamp. In turn these techniques employ carefully chosen materials, such as pigments imbued with power by the "solidified light" of quartz. To understand how he gives substance to the light-body of his vision, visionary images which embody his most profound experience of light, we must briefly visit their setting, the cave itself. Heir to a deep and rich artistic tradition, enlisting all his techniques and materials for this one purpose, he skillfully exploits its physical properties.

Flickering Light on Undulating Walls, and Pigment Power

When the shaman rekindled his lamp to paint after descending into the depths of the cave, he worked by a subdued, uneven light. Whether provided by a flickering torch or more likely a lamp burning fat or bone marrow, this light was integral to his projecting and contemplating of the cave paintings.

The images and compositions made by erratic light were, of course, also viewed by this same light, one alien to us in our age of day for night. Images morph in this light and the cave's changing sight-lines, given the crawling and craning necessary to view them. And everything moves with the undulating walls. This means that the painter and his paintings lived in the continual interaction of light and darkness. And it is this interchange that gives them life. As the viewer moved through the cave, his light isolated a painting by successive revelations. The animals "emerge," as Jane Brox says, "out of the darkness as you pass. Nothing stays still. Shadows nestle in the cavities; a flicker of light across pale protruding rock turns a hoof or raises a head." Figures fade from shadow into darkness and out again, or slide around corners. In such "soft, flickering light the animals take on an almost magical movement."[21] Or even more dramatically, the great Lascaux bulls at any moment stalk away into the shadows, just as they would from the hunter. Then just as suddenly they reappear.

Created and viewed by changing light, these animals owe their movement, life itself, to light. This act of painting, then, confronts us with the artist's deepest intuition of light. It is being itself, which is power to be appropriated. As in the cave so in vision, the light-body of the animal is revealed. It appears at the climax of the artist's descent into the blackness of the cave, the end of his journey into the "hallucinatory vortex" of visionary experience that David Lewis-Williams calls the "the mind in the cave." In both cases the painter returns with images. These did not *represent* visions, Lewis-Williams insists, but *were* visions. The painter's animals do not exist apart from the interplay of light and darkness generated by the quavering light of his lamp. They are brought into being by this particular light, a function of the cave setting. Imagine how powerful this was as animals suddenly loomed out of the darkness. Very likely the shaman saw literal bodies see-sawing on the fulcrum of the wall in and out of flickering light.

Just as the shaman exploited lamplight so he exploited the cave wall. He enlists both as partners in giving palpable form to the animal-of-power's light-body, projecting his own. Thus the cave wall participates in the painter's imagery, its varying textures, composition, and luminosity being an integral part of it. This suggests that he did not regard the cavern merely as a place to be decorated, but as a place believed to have special powers. His paintings dance with the "cave as participant." So the shaman took great pains to find walls that were exactly right for his painting. He chose them not only for their natural movement but for their natural luminosity, exploiting their "exceptional optical properties," which we'll return to. The wall, then, is not a passive, inert canvas any more than are the depicted beasts. The rock was apprehended, in fact, as a porous, "living membrane" between the everyday and the spiritual world, portal to a "realm inhabited by spirit animals." Just as some depicted animals, such as the bear, were mediators, so analogously was the cave wall. The animals of light passed through it in both directions between the natural and supernatural worlds.[22]

This parallels the movement of the shaman, of course, passing back and forth through his own skin on his vision journey, as he follows his "threads of light" between the internal and the external, this world and the other. As the shaman moves his lamp away from the wall, the animal depicted appears to retreat through the membrane of the cave wall. But as he moves the lamp closer, the animal seems to emerge out of the membrane toward us. Animal and man are partners in this dance, intimates. The shaman has the animal in his power as he moves the light away or toward the image. At the same time, the animal has him in its power because it disappears unless the shaman stays in a fixed position with the lamp. The undulating, luminous cave walls animated by flickering light give the shaman's images of the light-body life as well as power.

These skilled shaman-artists chose a site for painting only after many visits and tentative explorations, as we would expect because the living walls that host their images were not merely vehicles for paint like the typical bare canvas stretched across an artist's frame. They chose a specific wall with great care not so much to paint on but to partner with. Moreover, given their vision of light, which by means of their images they gave substance to, they were attracted to the light-reflecting and -dispersing properties of the rock in certain places. The brightly gleaming "calcite encrustation" covering nearly the entire wall of the Hall of the Bulls and of the Axial Gallery in the Lascaux cave, for example, has a unique "reflective power" that caught their attention. This unusually luminescent quality of some walls thus enabled the painter to realize his physical apprehension of light in all its tactile reality. The artist exploited this luminosity in the pigments he chose as well as in the methods he used to mix them.[23]

In giving substance to his intuition of a bodily light, the artist mixed pigments with clays that had a relatively high index of refraction, resulting in brilliantly reflective paints. They often included a white pigment, for example, composed of highly reflective mica and very fine-grained quartz, a mineral that was clearly precious to the shaman as a light-bearing substance. A natural conclusion, the painter having rubbed quartz stones together and discovered that they emit a flash of light (owing to a unique property called "triboluminescence"), which the shaman took as evidence of a supernatural power residing in the stone, just as entoptic flashes resided within him, that could be appropriated. Far from being merely paints on the Paleolithic painter's palette, then, his pigments possessed light-power in themselves.[24]

The Paleolithic artists embodied in their paintings the physical reality of light, its palpable presence, not just by selecting reflective "canvas" and paints, but by employing certain methods as well. Using limestone-rich cave water to mix their paints, which encouraged crystallization of the calcite on some cave walls, they applied pigments after re-crystallization. In this way, or by exploiting natural crystallization, they produced a glittering white by leaving a thin film of calcite unpainted. Given their visionary experience of light, the Paleolithic cave painters were understandably in love with gleam and glint. Reverence for brightness, epitomized by the light-body, is apparently lodged in the deepest layer of our consciousness. As a quality of the cave panels they found or prepared for painting and as an attribute of their pigments of power, brightness was a shamanistic obsession. All techniques used in numerous images and caves to optimize giving substance to the light-body of the animal mediator.[25] The mind in the cave is thus given substance.

The Visionary Image, and Interpenetrating Light-Darkness

Impressive as these light-exploiting (of lamp and wall) and light-bearing (in pigments) skills are, we must return to the shaman-artist's mission, for they were wholly subordinated to making concrete his fundamental experience of light as a light-body. Encountering this in the supernatural world and consolidating it in the natural world of the cave required animals of power, such as the bear. The carnality of light he knew in his identifying with these familiars who were guides to this light-body as well as instructors in the techniques of its incarnation on cave walls. At the same time, its visionary reality he knew in the fraught journey into the depths of his being and the cave. To attain and consolidate his fundamental experience of light's character—that the bodily apprehension of light and light itself cannot be separated—was his overriding task.

The shaman's visionary beasts, though having a "free-floating existence" on his walls, never give the impression of "floating in a vacuum," even though they are not referenced to the horizon or any landmark. This is because light itself constitutes their space. As they emerge from the membrane of the cave wall into the light, they take on being. Furthermore, they are often without hoofs or not standing, simply hanging, as in the Axial Gallery at Lascaux, just as we would expect from mental images that have been projected and fixed on the walls. These images, then, were not so much painted onto rock walls as "released from, or coaxed through, the living membrane," as David Lewis-Williams asserts, "that existed between the image-maker and the spirit world."[26] These animals could join herds of real animals, so animated do they seem. Nonetheless, they are spirit animals constituted by the interplay of light and darkness. As we'd expect of visionary animals, in their depiction they're severed from their context, their natural surroundings, and often presented in anatomically incorrect poses. The shaman's spirit animals are thus the means to vision and the vision itself.

So the Paleolithic artist was not compelled to depict but to give substance to visions of light braided with that of his ordinary experience. Again, we're reminded that these paintings rooted in the experience of light as a body did not represent visions but were visions. This is why, though these images of wild beasts are surprisingly familiar, feeling almost contemporary, we also experience them as strangely uncanny.

The painter goes beyond presenting the natural forms of animals and their supernatural light-bodies to indulging vision itself. In the "image explosion" of the Apse at Lascaux as one enters the cave, the shaman seemingly depicts passage into the light-body. We suspect, knowing the shaman's journey of ascent, that what he did on his descent into the Apse was a "way of passing into a virtual world beyond the wall" and "joining with the animal's vital spirits." We see here shimmering, incandescent, shifting patterns that were produced in the first stage of his trance (i.e., entoptic phenomena). And in experiencing this light-vision there is a palpable "sense of inhabiting a new corporeality in obbligato," an expression of the fundamentally carnal nature of the shaman's experience of light.[27]

But what of the beast images beyond the apse as consciously wrought pictures? These animals are at once given by the light and givers of the light, as the shaman experiences in his own body "light coming into being." Faced with the immense task of giving concrete reality to this light-body vision, the painter must use all means at his disposal, working with his partners, the cave theatre itself, as well as luminous pigments of power. Just as the shaman passes through his skin into the other world and his light-body, so these spirit animals experience the same in passing through the membrane of the cave into the light. In doing so, they reenact their role as mediators, as helpers on the shaman's journey into vision. On one level they are simply smeared pigment or charcoal traces employed to realize vision. On another, partnered with lamplight and cave wall, they embody the power that the shaman draws from animals. After all, we're stunned not by the guttering lamp itself or undulating wall, but by the animals of light moving in and out and across the walls.

To realize this vision, as we've seen, required great personal and technical resources, which include the artist's manipulation of light and darkness. The Chauvet paintings, oldest of the Paleolithic art, are primarily outlines that let the light shine in and out through the body of the animal. A stunning male ibex in The Megaloceros Gallery of Chauvet is outlined in black using the whitest, most reflective part of the wall, so light shines through the body, which is emphasized by the rightward sweep of the horns that divide the darker part of the wall out of which the animal emerges from the luminous part into which it merges.

The artist takes this carnal light-dark farther by interchanging black and white line engraving. He uses dark outlines, turning the luminosity of a wall to his advantage, as if literally creating the animal out of light, such as the lion on the left wall of the End Chamber. Elsewhere the effect is reversed, as the white outlines stand out against a yellowish wall. And the horses on the Panel of the Engraved Horses are composed of engraved white lines alone that precisely echo, no surprise, the earlier claw marks of a bear that accompany them.

We see that the twin brother of light has joined us. The animals of light we've met, such as the Bear of Chauvet and Bison of Altamira, do not come alive in the cave by light alone, but by darkness as well. Theirs is a "primal relationship." Not darkness apprehended as merely absence of light but as a being. For cave dark is not the darkness of the night or a room without a light, but a palpable substance, a body, just as is the light of the shaman's vision. Light and darkness cannot be separated in the theatre of the cave or in the act of painting because their dialectical interaction transmits into life its very significance.[28] The permeable membrane of the cave's wall between the painter and the spirit realm echoes the relation of darkness and light as well as above and below ground, the two realms of the Paleolithic world.

As a bird-man (like Raven) the shaman ascends into the sky to attain his vision; as a man-bear he descends into the cave to sustain his vision. The shaman ascends into the world of light to encounter the light-body, as we've seen. By taking on this body he attains power. But this is not sufficient in itself. He must descend into the earth, the theatre of darkness, as we've seen, to realize his vision by giving it substance. As a man, then, he lives in the ordinary world; as a shaman he lives in the "other" world of animals, the realm above, the sky, and in the nether world of darkness below, the cave.

Belonging to two worlds at once, light and darkness, he experiences his own perceiving-body as a composite of interpenetrating light and darkness. The shaman knows himself a sky-being and an earth-being, flying through the heavens and scrambling through the caves. And he is not alone in either realm, but in communion with his spirit-animal guide. And this echoes his ambiguous status in the community. On the one hand, the shaman stands outside the group because his concerns are with the inward, an inner world of vision rather than the outer world of the community.[29] On the other, he brings back power to the community as a healer, gained from his deeper experience of the character of light.

A power facilitated by his communion with spirit-animals above and below, as the bison of Niaux dramatically illustrates. His eye glittering in the darkness, the artist enters his own images not only as viewer but more important, as participant. Both darkness and light are required to give substance to this animal. The painter consciously added small details to a ripple in the rock's surface to bring the bison into being. The "distinctively humped dorsal line is clear," as David Lewis-Williams observes, only "when a light source is held to the left of and slightly below the image."[30] To similar expressive ends, the roughly contemporary painters of Lascaux consistently drew black figures on luminous surfaces to emphasize darkness/light contrast, or black drawing on luminous background.

They also used tonalities of light/dark to express their reciprocity. The famous Horse of Lascaux (a.k.a. "Chinese Horse") on one wall of the main gallery at Lascaux drawn in a decidedly non-naturalistic manner with tiny head and odd feet, yet giving the impression of bulk, owes this to the contrast of light and dark. The dense black of the nose and the porous black of the mane contrast with the belly's paleness conveyed by the unpainted, brilliant white calcite. Some fifteen thousand years earlier in Chauvet the painter used "smudging to produce shadow" and in one instance revealed the light at the edge of a drawing by "chiseling into the white rock surface."[31]

White interacting with shadow, such is the body of light and body of darkness that is the shaman-painter's composite body. This unconscious intuition of light and its brother as material beings is, as we've seen, Paleolithic peoples' core experience of light. Objectified by Raven in the shaman's tale, it is a profoundly physical light-dark. The shaman-painter who gives his intuition of light's body—his vision—substance on cave walls, reveals his relation to light by imaging animals of light in collaboration with living walls in cave-dark and flickering light.

Shifting from the earlier to the later Cro-Magnons, we move from the visionary to vision. That is, we shift in the Neolithic period from the shaman's vision of *being* light to the culture-hero's act of *seeing* the light-being, a sensuous seeing which he expresses by constructing chamber graves and megaliths. Hence I call this hero a "seer." Where the shaman-painter in an altered state of consciousness welcomes this body in vision, the seer-architect in a world of ordinary consciousness welcomes it in sight. Light shines out of the Paleolithic shaman's glittering eyes because it emanates from the inner body of light, whereas light enters the Neolithic seer's eyes because it emanates from the sun. As we would expect, this means that although both, being Cro-Magnons, share a fundamental bodily apprehension of light, the latter values it differently than the former. And this reveals a community whose concerns are very different from Raven's and a new aspect of light's character expressed in a new way.

NOTES

1. The Paleolithic, specifically, the Upper Paleolithic period (Lewis-Williams, *The Mind in the Cave*, 72).

2. Boas, "Tsimshian Mythology," 567. Widely distributed among the northwest Indians, these can be traced from the Asiatic side of the Bering Strait. Tales such as this no doubt have their origins in deep prehistoric times, as the Haida's Indian ancestors crossed the Strait some twenty thousand years ago.

3. Reid and Bringhurst, *The Raven Steals the Light*, 19–24. This tale was first reported by Franz Boas based on a text recorded by Henry W. Tate; Boas, "Tsimshian Mythology," 582. In an Eskimo version, "The Bringing of the Light by Raven," an apprentice shaman bests his elder, whose magic had failed to "bring back the light," by putting on a "raven coat" to steal the light, which he successfully brings back, breaking off "pieces of light" for all (van Over, *Sun Songs*, 83–85).

4. The philosopher Maurice Merleau-Ponty calls this the "knowing-body" (quoted by Vasseleu, *Textures of Light*, 46–47). That is, we know this "carnal light" not by contemplation but by our unconsciously sensing and interpreting body.

5. I follow Brian Fagan, in *Cro-Magnon*, in using this term for what are called in the scientific literature *Homo sapiens* or anatomically modern humans (Fagan, *Cro-Magnon*, xvii).

6. Pfeiffer, *The Creative Explosion*, 119; Lewis-Williams, *The Mind in the Cave*, 92, 95.

7. Lewis-Williams, *The Mind in the Cave*, 93–94. Steven J. Mithen describes this shift as being one from a "domain-specific mentality" to "cognitive fluidity," which he maintains is a "universal feature of modern minds, differentiating us from Early Humans" ("A Creative Explosion?" 169–70), making possible an "efflorescence of image-making" (Lewis-Williams, *The Mind in the Cave*, 192). Picasso, quoted by Nowell, "From a Paleolithic Art to Pleistocene Visual Cultures," 241.

8. My focus is primarily on the prehistoric experience of light in western Europe, Britain, and Ireland simply because we have the best evidence here recorded in the language of paint and architecture. For similar reasons, within the Paleolithic era my focus is primarily on cave painting, not because I subscribe to a narrow nineteenth-century definition of "art" as painting, but because it constitutes the best record of the experience of light in the era.

9. His images, in effect, are reified visions of light. Such "carnal" experience is rooted in the functioning of the shaman's nervous system as he entered an altered state of consciousness, a hallucinatory vortex that David Lewis-Williams calls the "mind in the cave." His paintings, that is, were external and internal (Lewis-Williams, *The Mind in the Cave*, 194, 204). I interpret the cave paintings of the Upper Paleolithic peoples as an expression of the "individual and collective experience" of light, just as David Lewis-Williams and David Pearce interpret them as the expression of "supernatural realities," which gives us, Donald Wiebe asserts, the "most coherent account of that data that we possess": namely, shamanism, as the leading Paleolithic archaeologist Jean Clottes concludes (*What is Paleolithic Art*, 172). The meanings given to this experience of light could well have been "spiritual" and shamanism "religion's first institutionalization" (Wiebe, "Roots in the Brain," 203, 206). But we do not know. What we know is the particular character of light embodied in cave paintings.

10. "Antiquity and ubiquity," Lewis-Williams, *The Mind in the Cave*, 206. For argument and counterargument regarding Lewis-Williams' cross-cultural definition of shamanism (*The Mind in the Cave*, 133), see Jolly, "On the Definition of Shamanism," 128 (Cecelia F. Klein, et al., reply); Layton, "Review: *The Mind in the Cave*," 422; Wiebe, "Roots in the Brain," 203–07; and Hayden and Villeneuve, "Review, *Lascaux*," 388. Based on Lewis-Williams' demonstration of its "antiquity and ubiquity," however, we can have confidence that "some form of shamanism was practiced by the hunter-gatherers of the west European Upper Paleolithic" (*Mind in the Cave*, 206) and safely dismiss P. A. Grand's charge that it is a "romantic notion" to see shamans or "sorcerers" on the walls of the cave or at work there (*Prehistoric Art*, 8).

11. "Harrowing journey," Brox, 9; "spirit helper . . . death," "portal to . . . status," Lewis-Williams (*The Mind in the Cave*, 265, pl 28). The shaman doesn't need an audience because he is his own audience, witness of his own transformation. So it is difficult to imagine anyone

other than the shaman himself making this image, and it's unlikely that other members of the tribe penetrated to the depths where this figure appears. Most images can plausibly be attributed to the shaman. These were probably done "either in a lightly altered state of consciousness or, after having returned to an alert state," as the shaman-artist "sought to re-create and control" his "fleeting visions" (Lewis-Williams, *Inside the Neolithic Mind*, 83). These mainly comprise pictures of animals not seen outside the cave. In short, it is likely that all shamans were artists, but not all artists were shamans.

12. "Concentric circles . . . brilliant dots," Lewis-Williams, *The Mind in the Cave*, 130; "the mind was strongly inclined to invent light," quoted by Pfeiffer, *The Creative Explosion*, 211. For confirmation of this neurological hypothesis, which Lewis-Williams proposes, see David Whitley, "Witnesses," p. 7; "most sacred place," Mario Ruspoli, *The Cave of Lascaux*, 146. Sacred or not, the profusion of images and signs here has been called an "art of latent excess" denoting an "unbridled zone" lying outside custom and doctrine, a zone of wholly personal, spiritual experience (Nechvatal, "Jean Baudrillard"); "threads of light" and "to climb the thread," Lewis-Williams, "Threads of Light," 129, 130.

13. "Experience of light coming into being," William Irwin Thompson, quoted by Vasseleu, *Textures of Light*, 45–46.

14. "Lightning," Eliade, *The Two and the One*, 23; "Every real . . . another man," Larsen, *The Shaman's Doorway*, 77; "mysterious light . . . metaphorically," Eliade, *The Two and the One*, 23.

15. "The light that shines . . . eyes" and "solidified light," Eliade, *The Two and the One*, 24–25.

16. "Wide-staring eyes," Pfeiffer, *The Creative Explosion*, 107.

17. Persistence in shamanistic cultures and shamans occupying the caves, Robert-Lamblin, "An Anthropological View," 204; bears as men, *Handbook of N.A. Indians*, 595.

18. This placement is undoubtedly deliberate. Not only does the stone sit in the middle of a spacious open area, but four lines of smaller animal skulls deliberately placed on the cave floor converge on the stone block, which supports the skull (Whitley, "Witnesses," 5). We don't know if this skull was the center of a cult or had any religious significance at all, but we can't escape its arresting presence.

19. And almost fifteen thousand years later we encounter a bison's light-body that dominates the ceiling of the Polychrome Chamber at Altamira (c. 14800 BP), exuding power. A great bulk, almost six feet long, yet he seems to be composed of light, as he floats in the "bright golden ocher" glow of the cave surface (Pérez-Seoane, "Technique, Individual Artists," 75). There can be little doubt that the animal has special significance, as its location was presumably a "highly respected place of creation" as there is no evidence of previous work being erased and the painted bulls in the Polychrome Chamber are all by a single artist (Pérez-Seoane, "Technique, Individual Artists," 65; single artist, 87).

20. "Entrails of the nether world," Lewis-Williams, *The Mind in the Cave*, 210. Thus it is unlikely that the cave paintings were done for purely aesthetic reasons, as most paintings were located in remote parts almost inaccessible or not easily accessible to viewers (Leroi-Gourhan, *Treasures of Prehistoric Art*, 33; Lewis-Williams, *The Mind in the Cave*, 149); and little evidence of practices of "sympathetic magic" to ensure the success of hunting (many of the animals depicted were not hunted), as the pioneering researcher in Paleolithic art, Abbé Breuil proposed (Leroi-Gourhan, *Treasures of Prehistoric Art*, 34; for survey see 33–35). For other speculations about purpose, see Leroi-Gourhan, *Treasures of Prehistoric Art*, 150; Hayden and Villeneuve, "Review, *Lascaux*," 385; Conkey, "Making Things Meaningful," 54, 62 n. 12. For a review of approaches, see Jean Clottes, *What is Paleolithic Art*, 19–24. It is clear that the cave art was informed by some version of shamanism, as Lewis-Williams demonstrates and Clottes confirms, *What is Paleolithic Art*, 172.

21. "Emerge . . . head," Brox, *Brilliant*, 9; "soft, flickering . . . movement," Giedion, *The Eternal Present*, 528. And see Clottes, *What is Paleolithic*, 102–03, 119–20.

22. "Cave as participant," Randall White quoted by Nowell, "From a Paleolithic Art," 242; "exceptional optical properties," Aujoulat, *Lascaux*, 257; "living membrane," Lewis-Williams, *The Mind in the Cave*, 161, 214, 221; porous hide, Giedion, *The Eternal Present*, 529.

23. "Reflective power," Aujoulat, *Lascaux*, 257; Curtis, *The Cave Painters*, 95; luminescent quality, Hayden and Villeneuve, "Review," 384.

24. "Triboluminescence," Lewis-Williams, *The Mind in the Cave*, 176–77; quartz in the pigment, Pérez-Seoane, "Technique, Individual Artists," 85.

25. Re-crystallization, Conkey, *Beyond Art*, 423. Reverence for brightness, epitomized by the light-body, is apparently lodged in the deepest layer of our consciousness. The anthropologist Nick Saunders speculates that the attraction to brilliance and shiny objects, common to all cultures, is part and parcel of the way our brains interpret visual stimuli. Glittering objects embodied the "power of light," a "cosmic power" (quoted by McDonald, "Pieces of Light," 43).

26. "Floating in a vacuum," Grand, 36 (and see Grand, *Prehistoric Art*, 36, and Giedion, *The Eternal Present*, 532); "free-floating existence" and "released from . . . spirit world," Lewis-Williams, *The Mind in the Cave*, 194, 199. We must remember that these mental images are embodiments as well of his experience of natural light. As Margaret W. Conkey reminds us, paraphrasing Svetlana Alpers, Paleolithic "imagery was created working from right *within* the mundane world. The 'artists' are not transcending but deeply involved *with* the world in which they live" (Conkey, "Making Things Meaningful," 49). Thus I'm mindful of the fact that not every single painting in the caves represents the subconscious objective of fixing other-world light (visionary) braided with this world's light (ordinary) on the walls. After all, the painters "lived so close to their prey that they were part of their quarry's lives," Fagan, *Cro-Magnon*, 209. As we would expect, then, a number of individual paintings, ranging from Chauvet (32000 BP) to Altamira (14800 BP) and Niaux (13000 BP), tell us much about late Ice Age animals. And we can't dismiss the possibility that some of the art may be purely decorative (Fagan, *Cro-Magnon*, 210). Yet while some familiar animals were easily visible at the front of the cave in its opening or vestibule, the animal "familiars" (mediators) lurked deep in the cave in its "profound darkness" (Fagan, *Cro-Magnon*, 211). Clearly, as Brian Fagan concludes, these images were "far more than art; they were objects of significance—images with potent ingredients of ocher and blood" painted on potent, living walls. And this notion is reinforced by modern evidence, as he points out, for "we only have to look at living hunter-gatherers to understand this," such as the Australian Aborigines and the San hunter-gatherers in southern Africa who continue a tradition of painting that is thousands of years old (*Cro-Magnon*, 149–51).

27. "Way of passing . . . vital spirits" and "sense of inhabiting . . . obbligato," Nechvatal, "Immersive Excess," 186, 188, 190.

28. Tymieniecka, *The Elemental Dialectic*, vii. Light, "owing to its primordiality as an elemental form, stands in a peculiar dialectic to darkness. It takes darkness as its complement, as the very means by which it may be known as light" (Ryba, "Elemental Forms," 18).

29. Campbell, *The Masks of God: Primitive Mythology*, 254.

30. On the bison of Niaux, see Lewis-Williams, *The Mind in the Cave*, plates 17–18; "distinctively humped dorsal line . . . below the image" (220).

31. Brilliant white calcite, Sandars, *Prehistoric Art*, 54; "smudging . . . rock surface," Robinson "The Cave of Chauvet-Pont-d'Arc."

Chapter Two

Planter Light of the Sun

The Neolithic Era

As Cro-Magnon peoples underwent a momentous cultural change from the life of hunters and gatherers to that of farmers and herders, who began to domesticate plants and animals in the Neolithic era, systems of belief arose that revolved around the sun and the dead. Soaring into the sky as a spirit-animal was now supplanted by riding into the earth on the sun's rays. What had been the "realm of spirit animals" in the Paleolithic caves becomes the realm of the "human departed" in the chambers of Neolithic graves. Shamanism persists to some degree, but the individual visionary of earlier times gives way to a ritual specialist. As the Paleolithic (c. 35000–10000 BP) fades to the Neolithic era (10000–6000 BP), that is, the "seer" emerges whose world is ordered by the settled farmer's cycle: night—day—night. And the earlier personal vision-quest becomes with the seer a social mission, a journey outward rather than inward as was the shaman's.[1]

SHAMAN TO SEER, BEING LIGHT TO SEEING LIGHT

On his visionary journey the Paleolithic shaman with the help of an animal mediator, as we have seen, underwent a transformation into a light-body. The hero of Neolithic society, however, leaves the light-body behind, abandoning the classic animal-shamanism of the Paleolithic period, but retains his seeing power.[2] He trades the powers of transformation for new ones, becoming a figure viewed as one who possesses the power of seeing into the nature of things, even beyond the grave. If the shaman was a technician in visionary practice, embarking at will on a personal journey of power to encounter the light-body, the seer is a specialist in ritual practice guiding his people, as we'll see, on a journey through the land of the dead. By this he achieves great authority and power within his community, his exalted status allowing him to dominate the religious life of his society. The shaman's personal power thus becomes the seer's political power.

Accordingly, the character of light in the seer's experience reveals different aspects that demand new expression. Light retains from the earlier period its independence of objects and attributes, such as heat, and remains a sensuous, "carnal" light apprehended fundamentally on the unconscious level. Assuming its body, however, as the shaman did the "light-body," now has no importance. Instead, a sensuous seeing that both celebrates and employs light in ritual comes to the fore. The shift from light's body as a material substance to the seer's body experiencing light brings a new focus in the later period. Neolithic society valued the act of seeing in itself, discovering that it "cannot be divorced from the carnality" of light.[3] This seeing derives from and depends on the seer's experience of light as part of him and not part of him at the same time. So as we'll discover, he takes pains to bring his body as well as society into concert with light. The later Cro-Magnons, then, knew light as something both internal and external. As *being* light yielded in importance over the millennia to *seeing* light.

As wandering groups of hunters settled into farming communities, the sun came into prominence as the giver of life, essential to cultivating plants. If the former period was a time of mastering animals, Cro-Magnon becoming the top predator, the latter was the time of mastering nature itself, cultivating plants and domesticating animals. Inevitably this revealed a dark side because the cycle of planting and harvesting depends on death. And this became personal. With the move from hunting to farming over many centuries, Neolithic groups found themselves occupying a more permanent location. Consequently, they developed an attachment to a specific place, the land they farmed. Participants in the cycle of sowing and reaping, governed by the sun, came to see the earth as naturally receiving their own bodies. And the land continued as individual lives ended. Attachment to the land became, then, attachment to their dead.

This change that came with the settled life accompanies a major shift from the sky orientation of the Paleolithic people to the sun of the Neolithic. After all, for the Cro-Magnon peoples of this era it is the hub of agricultural life, so not surprisingly their focus shifts. And now separated from visionary experience, light has a locus. We move with the seer from the undifferentiated light of the sky to the concentrated light of the sun manifest in its rays. In contrast to the Paleolithic shaman, then, the Neolithic seer placed great value on aligning himself with the source of light, the sun. And set apart as a specialist in this, an expert in seeing into the nature of things, he became the catalyst of cooperation by guiding the group in ritual participation. And in this he projects his relationship with light. His experience of the sun, then, included not only its centrality and its demand of observance but the sun's predictable cycle that offered a paradigm at hand for order. Differentiated as the sky is not, the sun also becomes the focus of the sky. It reigned supreme. Thus the sun, modeling harmony and hierarchy, took on great significance. Not surprisingly, in many myths the elite of society are descended from the sun, as the Inca and Aztec nobles believed.[4]

By projecting their experience of light, in short, Neolithic societies with their settled expanding population solved the problem of how to maintain stability. As cooperation in larger groups became an issue, they established an ordered hierarchy bound up with the prominence and movement of the sun. Although collective solidarity was sacrificed for stability, society gained the ability to organize individuals for large, collective projects. This enabled the communities of the period to build massive architectural monuments aligned with the sun and dedicated to ritual use. These monuments, as we would expect, grew out of the Neolithic belief system, one that reflected a familiar hierarchical cosmology—the heavens, the earth, and the underworld. In turn, this three-tiered cosmos justified belief in social hierarchy.[5]

The seer extends this cosmos into his monuments as well, such as Newgrange, a passage grave in Ireland, and Stonehenge in Britain. Just as Neolithic society was aligned with the seer, so these monuments were aligned with the sun, which presided over the sky-world of the seer's mythological cosmos. The shaman consolidated his light-vision in the underworld of the cave; the seer consolidates his in the middle world of his community. He does not transform himself but, instead, transmutes his experience of light into a built cosmology, monumental structures dedicated to the dead. Thus Paleolithic painters became Neolithic builders.[6]

The act of seeing for the seer involved touching the rays of the sun, to which these monuments and the rituals they enabled as well as the myths they echoed were dedicated. Myths, of course, had been told and retold for millennia. But monuments enact a new manifestation of light's character. They were completely new in the world, a distinguishing characteristic of the Neolithic world, as important as the advent of farming.[7] Such a constructed world had never before been seen in Europe. Simple long houses and ring cairns evolved into elaborate passage graves like Newgrange and massive stone circles like Stonehenge. As theatres of burial ritual, they participated in the alternation of the celestial bodies themselves—sun and moon. With this came the ability to predict, the power manifest in Neolithic culture of a seeing so intense that the future could be glimpsed. Such seeing combined observation with ritual identification. This essential involvement of the seer's body in the act brought about a new valuation of light and darkness.

The Paleolithic hunters we remember experienced the sun and sky as one, a hierophany, simply "the shining," a common notion in the earliest myths.

In contrast, the Neolithic planter experiences the sun as a separate entity. We see this in the culture of the Desana Indians of the Colombian Amazon who preserve some Neolithic cultural elements and imbue the sun with "yellow intention," yellow being "the color of the power of the Sun" by which the world was formed. The power of the sun's "intention" is a paradigm of the seer's power in his ritual.[8] Presiding within the megalith, he coerces the sun by means of this structure to take on physical form. It assumes a body in the presence of his people, becoming light observed. And this new relationship with the sun signals a change in the human experience of light. Where the shaman *becomes* the light-body, the seer *observes* the body of light.

Neolithic art evidences the new force of the seer's gaze, which ritually creates the sun-ray, by its depiction of the eyes themselves. In an early Neolithic near eastern village, Ain Ghazal, a number of clay statues have been found with eerily out-sized, staring eyes set far apart in the face (dating from 6750 to 6570 BCE), suggesting that their owner possessed the power of not only "seeing with his thoughts" as did the shaman but of seeing in this world the intention and body of light itself. The "eyeballs bulge slightly and are surrounded by marked oval ridges of black bitumen" to emphasize them. These artistic practices have no practical advantage, so the new quality accorded them connoting power in the world is very likely what the craftsman intended. Eyes that not only apprehend a new quality of light but demonstrate the power of that light.[9]

THE BRIGHTNESS, THE RAYS, AND QUARTZ LIGHT

The seer directs this new gaze outwardly rather than inwardly as with visions. And in his sensuous seeing experiences a new aspect of light's character—brightness and rays. The inner brightness of an earlier age, manifest in the shaman's eyes, is now externalized in celestial bodies—sun and moon—what the Maoris called the "chief eyes of heaven." As a consequence, this seeing is a public act performed in the ritual space of monuments aligned with sun and moon. It remains a sensuous seeing because it involves the sun's body according to, as the Desana Indians say, its "yellow intention." Yet such seeing does not necessarily involve sun worship, a later development and surprisingly rare practice in the world.[10]

Although the "eye" of the sun now asserts itself as light's source, the seer continues to experience a light independent of its source, as did the shaman. The fundamental prehistoric experience of light has not changed. What has changed as hunter evolved into planter culture is the valuation of certain aspects of this experience, which is driven by the fact that the Neolithic seer is no longer a visionary participant becoming the light, but a ritual observer identifying publicly with the light and being, in a ritual sense, by power of prediction its creator. The seer does this, paradoxically, in order to observe and control the light-body via ritual, to socialize a sensuous light in effect.

Just as the relation of light and darkness took on a new meaning, allied with life and death, so does brightness. And for the first time in our history, humans distinguish and value rays. Brightness for Paleolithic painters was a means to an end. In fixing the light-body on their cave walls, they exploited calcite-luminous walls and quartz-infused pigments. They were thus obsessed with bright substance. And brightness was a manifestation of power, as it is for Neolithic builders. But later Cro-Magnons observed and valued brightness in itself. It was not simply something to be exploited but an integral part of the character of light as they experienced it.

In ritual practice brightness was inseparable from rays as embodied in monuments and experienced notably in the Neolithic innovation of a ritual burial. The dead became an element in their cosmic scheme, positioning the ritual specialist seer and his followers at the "point of transition" between worlds and thus in a position of power. As the chief celestial body, the sun is the figure of the seer himself who dominates his community as the sun dominates the sky. Publicly identifying at a burial with the shaft of light rays, which at specified times pierce the sacred space that he presides over, and drawing on a spiritual power manifest in the brightness of materials used to construct these bolstered his status.

In the surviving images we've seen from the earliest myths we find the expression of an ineffable experience of brightness and the sun's rays. And from myth to monument was but a step. The Neolithic belief system of a triple-tiered cosmos traversed by the dead who embarked on their journey in ritual burial drove the building of monuments dedicated to harnessing the sacred rays of sun and moon on their behalf. The Paleolithic painter made this journey, a type of death, in life; the Neolithic builder facilitated this journey, by the acts of the seer on behalf of the living, in death.

Because the essence of the sun and moon revealed itself as brightness, bright things on earth like minerals revealed their sacredness in their brightness.[11] Hence the seer used quartz crystals in his ritual practices, doubtlessly supplemented with personal adornments, such as a bright disk of obsidian hanging from his neck. Possession of light-bearing objects thus conferred power and status. The shaman painter experienced the glitter of his calcite-encrusted walls as revelation of light's body. Mineral brightness enabled him to give substance to the light-body. For the seer, however, brightness was a literal manifestation of the sacred realm, a spiritual world parallel with but distinct from his own. This world was "inhabited by bright spirit-beings," commonly thought to be forces in the world.[12] A reality in itself, brightness makes its presence known.

The seer projects his experience of brightness, which quartz incarnates, as power in life and death. In life it becomes political and in death religious power. He controls the spirit-beings of brightness by ritual practices involving the dead who occupy the threshold between worlds. His is a cosmology, in fact, sustained by the dead who mediate between life and death. In the former time, the shaman used quartz to usher himself past the threshold between natural and supernatural worlds into the inner world of vision; in the later age, the seer as ritual specialist filling a public role uses it to guide others over a threshold into the world of the tomb, room of the dead, the literal underworld. Both derive power from the spirit-beings of brightness that inhabit quartz. But where the shaman ingested quartz to enable his personal journey, whose object was the light-body, the seer built with quartz to enable the ritual progress whose object was welcoming the light-shaft.

To grasp just how significant quartz as a light-bearing substance that drew its power from the sun was in the Neolithic experience, we have only to visit the principal passage grave of the age—Newgrange (c. 3000 BCE).[13] The first thing that strikes us on approaching Newgrange is its brilliant white façade of quartz glittering in the Irish sunlight of county Meath near the River Boyne. Furthermore, the entire structure lies on a layer of stone composed of white quartz, the predominant material, and granite boulders, which extend the length of the passage tomb and outwards some fifty meters to each side of its entrance. This wide wedge of stone, thickest at the entrance, required huge quantities of quartz that had to be brought over forty miles by boat from the Wicklow Mountains in the south, further than any other material used. And this was not an ordinary variety but special quartz containing distinctive large crystals of white mica (muscovite) that enhanced its glitter. So important was this quartz to the builders that they conveyed it at great expense, involving an army of workers who had to quarry and transport it first by sea and then up the Boyne River, a truly amazing feat.[14]

Entering the grave, we notice that not only was quartz essential to construction, but the builders used it to indicate a ritual or sacred space. They laid it out strategically at the entrance itself. Quartz marked the threshold between life and death, being thus essential to the burial ritual that transpired in the completed passage grave. This use of quartz was not confined to Newgrange but appears widely throughout the British Isles. In guiding participants across the threshold of quartz, the seer enacted publicly a transition that the shaman had before him undergone privately. This was a public performance of the buried one's passage from life to death. Quartz, because it comes from the earth below and catches the sun above, acts as a mediator between the two realms. The Desana of South America to this day believe that the way in which quartz glitters with brightness or shimmers almost iridescent signifies "dynamic change," and they bury their dead in modified canoes that mimic the hexagonal shape of a quartz crystal.[15]

TRAPPING THE RAYS—NEWGRANGE AND STONEHENGE

While brightness and its materialization in quartz was an extension of a belief that survived from earliest times, rays were a new aspect of light's character. The seer inherited brightness, which he invested with a resonance it had never carried before, but he "discovered" rays. Not new in the world, of course, but these took on a new importance, becoming a new dimension of prehistoric light. His Neolithic megaliths were dedicated to capturing and controlling the sun's rays for ritual purposes.

Experiencing light as fundamentally lively rays emitted by the eye of the sun and received by the seer's eye could occur only with a shift from the sensuous being of light to the sensuous seeing of light's body. A shift prompted by Neolithic people coming into a new awareness of the day-night-day cycle of farming, as we've seen, and feeling the concomitant pressure of increased population and need for stability. Control of these cyclical, impersonal forces took center stage, as they had not in small hunter-gatherer groups. Likewise, the sun's position in the sky moved to the center of consciousness. And now a common phenomenon of sunlight demanded attention, such as rays spearing the earth.

This experience of another dimension of light's character, the Neolithic people relating to light in a new way, the seer makes known in his art. Carved or picked into many of his monument stones is a "rayed sun," the motif most often found in passage grave art. More significant, his relation to the sun finds a material expression in his monumental architecture itself. Neolithic megaliths were constructed as machines to trap light and harness it for the benefit of the community. And once he can capture the rays of the sun or moon in his megalithic monument, a light-trap precisely aligned with them, he can track and predict the appearance of celestial bodies, enhancing his power and status.[16]

Yet his monuments are not, in fact, astronomical observatories. He doesn't need them to determine when to plant and harvest, or even to predict the solstices, assuming farmers and herders cared about these. To track the rising or setting of a celestial body as does the astronomer is not the seer's purpose. So why did the Neolithic builders, with their technical knowledge of "levers, fulcrums, foundations, sheerlegs, slings, and ropes" as well as of tides and astronomical movements, to say nothing of surveying and quarrying, use these skills to build Newgrange and Stonehenge?[17]

The answer is that their careful architectural alignments with the sun and moon had, as the archeologist-astronomer Clive Ruggles insists, a "symbolic function." But "symbolic" must be understood in the context of Neolithic burial rituals and beliefs. In this context its function is not an act of abstraction but rather enactment. The ray of light assumes a role on stage. The seer seeks to experience its light-giving character in a particular position. Thus the megalith does not point beyond itself as a symbol but embodies in its materials and structure the transition from life to death, from the natural to the supernatural world. As the seer leads participants in their burial ritual, his monument becomes a physical emblem, an expression of the Neolithic experience of light rays that encompasses its mediating character. It is this distinctively sensual "astronomy" that the seer incorporated into his ritual practice.[18] So the rays the seer catches in his megalithic light-trap are not abstract notions but sensuous beings, players in a drama. Like Neolithic time and space in sacred ritual and its theatre, respectively, light-rays carry an emotional charge. Light from the sun or moon is not simply observed; it is intensely seen, touched with the eyes and felt as palpable being.

The Ray-Gate in a Passage Grave, Newgrange

Guided by the seer into Newgrange just before sunrise at the winter solstice, the Neolithic participant in the burial of a loved one would, on looking into a recess on the right, see a rayed circle, a sun sign chiseled on a stone basin holding human remains. The sign faces down the main passage towards a midwinter sunrise. Presently, a "thin thread of sunlight" begins "striking in along the passage," as the anthropologist Claire O'Kelly describes it, "until the dark of the chamber begins to disperse and more and more of it becomes visible as the sun rises and the light strengthens" (Figure 2.1). And looking outward towards the entrance, he sees "the ball of the sun dramatically framed in the slit of the roof-box," a rectangular opening just above it.[19]

This brief time is the only period in the entire year when daylight dominates the darkness of the tomb. After this seventeen-minute drama of sun rays penetrating some nineteen meters to the back of the chamber has passed, the participant approaches the ingenious sun-gate at the end of the main passage of the grave, the roof-box itself, which rests on the first two roof stones of the passage. Resembling a lintel, it is cunningly constructed with the inner part opening into the passage. Through this aperture the taut thread of sunlight had passed into the tomb's innermost chamber. On a large slab above it resting on the side walls of the box are engraved circular motifs. Each has a dot in the center with six lines radiating from it like rays. Also, the Newgrange builder designed the box to allow sun into the grave after the dead had been placed inside and the entrance permanently blocked with a stone weighing nearly a ton. Fittingly, the etymology of the Gaelic name for Newgrange, *An Uamh Greine*, is "Cave of the Sun."[20]

Clearly, Newgrange is not an observatory. It functioned primarily as a tomb, of course, but more important, incorporated a ritual of light essential to the seer's burial ceremony.[21] Interpreting his experience of the sun-rays as power, he appropriates this power in the ritual light-act. And as the rays of the sun creep along the floor steadily deeper, deeper, it seems to those gath-

Figure 2.1. Light entering the passage at Newgrange at sunrise on the winter solstice. ***Michael Fox, BoyneValleyTours.com.***

ered that the seer controls them, finally guiding them to touch the sun sign on the pan at the end of the passage that held the remains of the deceased. These rays of promise constituted the burial ritual. Neolithic participants in the ceremony were prepared for these by a brilliant, sparkling quartz façade, as we've seen, and then ushered into the netherworld of the dead over a quartz threshold. The seer's rays are life itself beyond the darkness that is death. But the brightness alive in quartz brings them to darkness and out of darkness as well. For only the brightness of mediating quartz can bring them back from the world of the dead.

The Corridor of Rays in Stonehenge

The ray character of light bound up with burial ritual remained remarkably stable over a long period of time. It was a witness to the seer's new relationship with light rooted in a new experience. And monuments built during this period tell us that this relationship was vitally important to the culture. About a thousand years after Newgrange was constructed, the first phase of Stonehenge was completed. Beginning about 2550 BCE, the structure was centered on the rays of the sun as it occupied a particular position in the sky at a particular time. The importance of the seer's new experience of light to the culture could not be clearer. The construction of such a monument dedicated to the new, ritualized aspect of light's character required many centuries and vast resources.

Monument to rays and the seer's special relation with them, Stonehenge dominated the Neolithic culture of western Europe for about a thousand years until it was finally abandoned (c. 1600 BCE). No one knows why. The circle of Stonehenge, a shape persisting over the course of its fifteen hundred-year construction and possibly an emblem of the sun's orb, was not only the precinct of the sun, but the "hub" about which Neolithic society "slowly revolved." And the celebration of a particular aspect of light's character was its pivot. The discovery and valuing of rays came together with the ability to communicate this on a large scale. By the mechanism of "ritualized architecture," a Wessex tradition for some three thousand years that includes Newgrange as well as Stonehenge, the seer's experience of light as rays became a communal one.[22]

The importance of the rays in their ritual theatre becomes clear when we realize the immense resources that were devoted to them. Hauling just one of the seventy-seven sarsens, the largest stones of Stonehenge, each weighing some twenty-six dragging tons, twenty miles from the Marlborough Downs required the labor of hundreds of the seer's neighbors and those from surrounding communities. Furthermore, the story of this construction, as the design of Stonehenge evolved over the centuries, is the story of an increased concern with rays. As the builders made successive changes, its ray-capturing purpose comes into sharper focus. With the advent of more intense farming, as labor on a large scale became available, the last stage of building began (c. 2550 BCE). This new surge of effort included bringing bluestones in to replace the earlier timbers and, probably, erecting the "Heel Stone" sarsen, over which the sun rises on the summer solstice.[23]

Significantly, in this late stage when stone replaced wood as the monument achieved preeminence, it was shifted "several degrees in its axis to bring this into line with summer solstice sunrise to the north-east and winter solstice sunset in the south-west." An adjustment of particular value as well, no doubt, was the alignment of the "Altar Stone with the direction of midwinter sunset in the south-west." Even more important in tuning Stonehenge as a ray-capturing instrument, the entrance stones were lined up with the Heel Stone and its companion in order to produce a "'corridor' down which the light of the rising midsummer sun would have shined directly into the interior of the monument."[24]

With this corridor Stonehenge is brought to perfection as a ritual space dedicated to attending to the rays of the sun. Its architecture was designed to dramatize this midsummer event and give it the maximum emotional charge. The seer draws on all his community's resources to convey the power of his core experience of light. To witness rays streaming within the precinct of the sun while the seer led a burial ritual for a loved one would have been undoubtedly an overwhelming experience. Accordingly, Stonehenge retained its importance in Wessex as a ceremonial focus for about a thousand years.

Central to the life of Stonehenge, then, was the one responsible for bringing the light down the corridor at a certain time every year. This intimate knowledge of the sun's character manifest in rays that produced the annual drama of light at sunrise and sunset placed the seer in a powerful position. To the community it was magic. The seer promised to welcome the midsummer rays on a certain day. And they appeared as predicted. As a result, the seer was no doubt highly honored, a hero of sorts, having an intimate relationship with the sun.[25] Magic, yes, but magic with a purpose, the drama of rays plays out in the ritual context of a burial ceremony. The seer apprehends and participates in light that binds the natural and supernatural together, linking the world below with the world above. And by this gives public expression to a personal experience of light.

The seer did not employ Stonehenge, then, as an astronomical instrument to explore the heavens, but as an architectural machine to seize the sun in a particular position. We remember that centuries earlier an elaborate burial chamber, Newgrange, was constructed to usher in the sun at certain times of the year. Similarly, the "trapping of light above the Altar Stone" of Stonehenge no doubt indicates a "ritual of illumination" centering not on the sun but on its light.[26] And this ritual of light, of continued life, experienced apart from its source, was bound up with the ritual of burial, death.

At the moment of the sun's capture, rays could be employed for human ends celebrated in specific rituals that provided consolation and hope in encountering death. The world of life and the world of death are given continuity in the Stonehenge drama of light. And the ray plays the lead role in this drama. As in Newgrange, rays are bound up here with burial ritual, cremated human bones having been found in Stonehenge as well as the surrounding area, including the remains of individuals from Wales and Germany. The seer understood death to be a passage from one realm to another, a passage facilitated by his presiding at the ritual burials that took place within Stonehenge, which was constructed to recapitulate this passage. As he moves through the three main circles of Stonehenge into its center where cremated bones lie, the realm of the sacred, the seer moves into the netherworld. So death and the monument's structure itself were inextricably bound up together.[27]

Yet this new light the seer experiences, which informs the ritual theatre, insists on its own presence. The ray is the protagonist, not the seer. In contrast to Newgrange, Stonehenge lies open to the sun. Its main avenue roughly targets the northerly extreme of the rising sun, framed by upright stones placed at the inner portal and offering a spectacular sight as "its rising rosy radiance glinted off the top of the Heel Stone," present from the earliest phases of construction.[28] At the same time, rays of light orchestrated by the architecture and the seer pass down the corridor.

Both Newgrange and Stonehenge, then, served primarily as theatres for the seer's burial rituals in which rays of the sun performed, orchestrated not by an astronomer but a ritual guide. This entailed not merely observing but "touching" light with the eyes, the seer's intense and intimate looking. And this new relation to light's character in the Neolithic together with his culture's focus on the dead demanded a new expression, the megalithic light-traps—the ray-gate of Newgrange and the ray-corridor at Stonehenge. As rays were ushered in, these traps also served as the means to dramatize his relationship with light publicly. Within his stone monuments, after all, he stood uniquely on the threshold, marked by quartz with its spirit-attesting brightness, between worlds in passing from the natural to the supernatural realm, from life to death. A drama of burial, certainly, but it also likely recapitulates a drama of the seer's personal encounter with light.

So where the shaman's cave hosted the light's bright body, the seer's megaliths snagged the light's rays. Yet no fundamental break exists between what I've called the "hunter light of the sky" and "planter light of the sun." Both the Paleolithic shaman and Neolithic seer experience a carnal light, one intimately related to the body, and both experience this light as independent of its source, sun or moon, even as they were no doubt aware of its source. Such is the core prehistoric experience of light regardless of the earlier and later prehistoric cultures having different needs and thus demanding different expressions. The shaman, in order to heal, assumed the light-body above in the sky and confirmed it below in the cave, giving it substance in paint; the seer, in order to mediate, embraced the sun's light-ray seized in transit by his megalith.

But new dimensions of light's character reveal themselves. The valuation of brightness did change. A means to an end in the shaman's experience, it was an end in and of itself in the seer's appropriation of its power. And the Neolithic seer discovered rays, valuing their sensuous independent presence. Paleolithic painters becoming Neolithic builders signaled a shift not so much in the prehistoric experience of light as in the meanings they attached to it. Hunter light meant the power to heal and ensure the return of migrating food animals every year. This delivered the community from the darkness of helplessness, threatened by illness and starvation. Planter light meant the power to cross the threshold safely from life to death, from light to darkness.

Likewise, light and darkness were brothers from the beginning, but their relationship was revalued with the advent of settled farm life. The shaman encounters darkness as an equal; the seer manipulates darkness as a tool, making his light-trap possible. It is a gate he opens to let in the rays, dramatizing an experience of light and darkness that emerged from a new cognizance of the day-night-day and seasonal cycles, which came with farming. The shaman focuses on the present, then, the seer on the future. Yet both share a basic experience at an intuitive level of the same independent and sensual light.

This continuity breaks in the next chapter. There we'll meet the priest, which marks a fundamental change in the experience of light. With the rise of city-states in the Egyptian Classical Age, we not only leave the shaman and seer, but encounter a radically new valuation of light's character in the realm of the king who is inseparable from a new class of ritual specialists—the priesthood. Certainly, religion figured in prehistoric Paleolithic and Neolithic cultures, but the emphasis was on religious *experience* enacted, respectively, in personal and public ritual.

In the Prehistoric Age, the sun and moon were not gods. The alignments on the sun as well as the moon at Stonehenge bear this out. They point to the fact that the seer and his community did not primarily value the sun and moon in themselves as objects of worship but rather their light as a substance of power.[29] All this changes in the next age. Here we move from a preoccupation with the "common" and "simple" character of light in Reverend Hunter's terms, to the celebration of its "glorious" and "mysterious" aspects later projected onto gods.

NOTES

1. "Realm of spirit . . . departed," Lewis-Williams, *Inside the Neolithic Mind*, 84. The Neolithic begins in the Mideast about 9000 BCE (Mithen, *After the Ice*), in Europe after the last Ice Age about 8000 BCE, and in Ireland about 3000 BCE (Lewis-Williams), though some maintain that here the early period was 4250–3250 BCE and the late 3250–2250 BCE (Ruggles, *Astronomy*). Told by a tribe of planters and herders in the Banks Islands of the South Pacific, an ancient tale featuring Qat, the most prominent figure in their mythology, captures this shift to Neolithic. In contrast to Raven's quest for light that delivers him from darkness, Qat's quest is for darkness to deliver his brothers from light. It is the materiality of night that strikes us in this tale, as it parallels that of light in the raven tale. However, Qat's journey is precipitated by an act of seeing, the brothers complaining that "It's too light." This myth was originally collected by R. H. Codrington (270–72). I use Sproul's retelling (332–33) from Maria Leach, *The Beginning*, 178–81.

2. Lewis-Williams, *Inside*, 148. We should keep in mind, nonetheless, that a number of early farmer-herder cultures include shamanistic aspects (Jolly, "On the Definition of Shamanism," 127). And the earlier and later societies overlapped for at least four millennia (Fagan, *Cro-Magnon*, 7). Nonetheless, anthropologists are uneasy about the use of "shaman" in "incipient agricultural societies" (Lewis-Williams, *Inside*, 125). As with the shaman whose experience of light we regarded as "representative" of the Paleolithic era, I treat the seer's experience as representative of the Neolithic. We can proceed with some confidence founded on the surprising continuity of cave painting as well as shamanism over millennia in the earlier era and of monument building over a millennium in the later era. The ubiquity of the "shaman-bird," for example, known from myths as well as from surviving worldwide ritual practices for his ability to fly "to the heavens, . . . to the moon and sun, to distant stars and back again" (Joan Halifax quoted by Lewis-Williams, *Inside*, 67). Similarly, the ubiquity of megaliths throughout the Middle East and Europe is well known.

But how truly "representative" of Cro-Magnons the shaman or the seer was we cannot ever know, of course, because we only have the testimony of their paintings and buildings, respectively. And these were made by individuals set apart from their community, at least to some degree, by their apparent authority. The shaman and seer, though extraordinary members no doubt of their respective communities, dramatize the culture's core experience. We can have some confidence, then, that they embody the values accorded to light in prehistoric societies.

3. "Cannot be divorced from the carnality," Vasseleu, *Textures of Light*, 45.

4. The shift from sky to sun (Armstrong, *Short History*, 57) is recorded in ancient myth as well. For where we have evidence in the mythological world of the existence of gods, sky gods tended to become either fertility gods or sun gods (Eliade, *Patterns*, 110, 139). Ceramic statues in Jericho and Cernavoda as well as megaliths centered on burial in Britain and Ireland attest this. For these megaliths are commonly laid out to align with the transit of the sun. And passage graves designed to capture the sun, as we'll see, exhibit reverence for the dead as well as anxiety about death. Light honors the dead and sustains them in their passage into darkness, which for Neolithic people was the world of the dead inseparable from the natural world.

5. Nomadic Paleolithic groups were relatively small and the demands of cooperation modest, requiring no hierarchy. While the smaller hunter-gatherer tribe was mostly egalitarian it appears, the larger Neolithic planter community generally admitted more distinctions in wealth, status, and power. And now the problem of enforcing cooperation in a community where inequality was inevitable became acute (Campbell, *The Masks of God: Primitive*, 241, 468; McNeil, *The Rise of the West*, 25; Bellah, *Religion in Human Evolution*, 175, 214, 262).

6. In the seer's ritual passage descending from his middle world into a megalithic grave to encounter light, as we'll see, and then ascending from this underworld to the heights of its mound, thought to be the heavens, the seer recapitulates the shaman's old inner journey (Lewis-Williams, *Inside*, 88). We see remnants of this Neolithic cosmos in the houses of the Barasana Indians who live in the Vaupés region of Colombia. Their *maloca*, a long house, embodies a three-tiered cosmos: "the roof is the sky," as they insist, "the house posts are the mountains that support the sky, and floor-space is the earth," beneath which is the netherworld (Stephen Hugh-Jones, quoted by Lewis-Williams, *Inside*, 90). Significantly, a small vertical post called "the Sun's post," placed in the roof of this house, is supremely important to these people. "Every day at noon, the sun's rays re-establish it as they fall directly to earth," an ethnologist reports, and this is repeated in their ritual. Furthermore, the major east-west horizontal beam on which the Sun's post rests is called "the Sun's path" (Stephen Hugh-Jones, quoted by Lewis-Williams, *Inside*, 90–91). So the sun is more important than the sky, having achieved its dominance in the Neolithic.

A Zuni myth tells the story of how this came about when in "the Beginning of Newness," Awonawilona, Maker and Container of All, "conceived within himself and thought outward in space, whereby mists of increase, . . . were evolved and uplifted. Thus, by means of his innate knowledge, the All-container made himself in person and form of the Sun whom we hold to be our father. . . . With his appearance came the brightening of the spaces with light. . . . With his substance of flesh outdrawn from the surface of his person, the Sun-father formed the seed-stuff of twain worlds. . . . Thereupon the Earth-mother repulsed the Sky-father, . . . thus separating from the Sky-father." Her progeny then cry out, "how shall our children when brought forth, know one place from another, even by the white light of the Sun-father?" (Thompson, *Tales*, 17, 37).

7. Bradley, "Architecture," 228. Contemplating the alternation of day and night, light and darkness, as any farmer must, moves space and time to the forefront of consciousness, both rooted in the subconscious apprehension of light. Such contemplation brings about "an awareness of space as space," the philosopher Ernst Cassirer maintains, where the sun rises and sets, west as opposed to east; and the "articulation of time, marking it, segmenting it into hours, seasons, and years" (Cassirer, *The Philosophy of Symbolic Forms*, II, 107).

8. "The shining," Eliade, *Patterns*, 40; "The Sun created . . . world from it," quoted by Classen, "Creation by Sound/Creation by Light," 246–47. Mircea Eliade points out that the example of the Desana Indians illustrates how a specific experience of sunlight can "valorize the light-experience and integrate the hallucinatory visions in the structures of an ecstatic universe," which "enriches their traditional religious system" (Eliade, "Spirit, Light, and Seed,"

29–30). And this system in turn valorizes subsequent experience of light. It is this dialectic that I trace in the chapters that follow.

9. "Eyeballs bulge . . . ridges of black bitumen," Lewis-Williams, *Inside*, 72–74, 76. Such exaggerated, alien eyes can be seen in a large sculpted pottery head from Predionica, Serbia, Yugoslavia (c. 4000–3000 BCE; Sandars, *Prehistoric Art*, illustration 95), whose intensity is heightened by their size and slant on a bold triangular face. Also, the late Neolithic engraving filled in with white stippling on a red clay pot from Voznesenskoye, Amursk district, Khabarovsk territory (3000–2000 BCE), exaggerates the paisley-shaped eyes with their stems curving out to the right and left, making them wholly out of scale with the rest of the face (Okladnikov, *Art of The Amur*, plate 19). The beautiful curves of this "anthropomorphous" image depicting large, stylized, penetrating eyes echoed in the heart-shaped outline of the face make them otherworldly rather than alien. Another face resembling an ape mask incised on a basalt cliff (4000–3000 BCE; Cheremetyevo, Viazemsky district, Khabarovsk territory) with eyes as the prominent feature is complemented by lines bursting from it as radiance (Okladnikov, *Art of The Amur*, plate 14). The Wiradjeri of Australia believe that a supernatural being known as Baiame "sings a piece of quartz crystal into their foreheads so that they will have X-ray vision." Such vision manifests itself in a common Neolithic artistic practice known as "X-ray" style, which typically reveals the inner organs of an animal's body. Because they are neither accurately depicted nor placed in the body as they would be for a hunter's purposes, it's probable that this style was developed in order to reveal the power of seeing itself ("sings a piece of quartz crystal . . . X-ray vision," Lewis-Williams, *Inside*, 257).

This style is a vestige of earlier shamanism that persists in the overlap between shaman and seer societies. The seers inherited powers of "vision" though fierce animals and "visionary" power receded as the community came into the foreground. Game animals were less important, of course, to a society that primarily relied on domesticated plants and animals. The X-ray style often seems not to be depicting specific organs, but the very act of seeing the animal's spirit. Emphasis is on the power of the shaman's act of seeing. The Indian artists of the northwest who tell the tale of Raven stealing the light employed this style as well. In their variant, the joints of a body are represented as eyes or concentric circles (Haberland, *The Art of North America*, 66). Like the rock carvings of the San, the emphasis of X-ray style is not on a hunter's map of the organs or the skeleton, however, which in any case often don't coincide with actual placement, but on the power of the shaman's seeing.

Clearly related to late Aurignacian art, such as that of Chauvet cave in western Europe (Lommel, *Prehistoric and Primitive Man*, 44), a rock engraving in Norway (c. 5000–2000 BCE) of elks mating suggests a map of the animal's inner organs (Sandars, *Prehistoric Art*, 81, figure 25). But given the symbolic power of the elk within this Neolithic society and the beauty of its lines, this cannot be interpreted simply as a map guiding a hunter to the heart, but something more significant and mysterious. It reveals the residual power of the shaman's gaze, the power of the light from his eyes.

10. "Chief eyes of heaven," Eliade, *Patterns*, 124.

11. Brightness as an aspect of light's character appears to be a world discovery, capturing seers from Europe to the New World to China. This ancient love, which continues from the Paleolithic, perhaps explains why thousands of years later a shining figure like El Dorado in Mexico went about "covered in fine gold dust," as Gonzalo Fernándo de Oviedo reported in 1514, "like powdered sunlight." And Aztec myths tell us that mica was "excreted by the moon, symbolized cosmic forces, and possessed a soft, buoyant light." Brightness possessed significance owing to the fact that throughout the Americas celestial bodies as well as glittering earthly objects were conceived as essentially "materializations" of light (Saunders, *Prehistoric Art*, 227, 232, 244). For this reason, amber was highly prized in the Neolithic culture of Europe and traded widely. It was valued for its light-bearing properties and highly polished to emphasize its translucent glow and shininess. Though carved amber objects often display much ornamentation, it is subordinated to their glow, reinforced by the fact that it seems to serve no functional purpose. A sculpted horse from Germany, for example, consists of rounded, globular forms, including the head and an added hump at the shoulders to exploit a reflective light, the shine being drawn from the amber's inner light (see illustrations, Torbrügge, 53 and 55). In China's Neolithic period, jade served the same purpose as amber and possessed the same

attributes and power, being associated with an earlier shamanism. It endowed the bearer with the power to ascend from earth with the assistance of his animal helpers (Chang, "China on the Eve," 63).

12. "Inhabited by bright spirit-beings," Saunders, *Prehistoric Art*, 226.

13. Twohig, *The Megalithic Art*, 11–12; Stout, *Newgrange*, 18. Ruggles dates the later period c. 3250–2250 BCE (*Astronomy in Prehistoric*, 84–85). But Lewis-Williams leans to a later beginning, 3600–3100 BCE (*Inside*, 204). Just as there was an amazing concentration of Paleolithic cave art in western France and northern Spain, so later there flourished an astounding concentration of Neolithic monuments within the Bend of the Boyne in Ireland, where Neolithic people built a range of megalithic passage tombs.

14. Michael J. O'Kelly, *Early Ireland*, 68; Eriksen, "The Great Mound of Newgrange," 255–56; Cooney, "Newgrange," 702; Lewis-Williams, *Inside*, 227. Lewis-Williams believes Newgrange to be a genuine reconstruction rather than figment of the modern-day restorer's imagination because no quartz was needed to be imported for the renovation work; all was found piled at the base of the grave (226–27).

15. In earlier monuments oriented to the moon, such as the short stone rows at Glengorm and at Ardnacross on the Isle of Mull, they scattered hundreds of jagged quartz fragments among the standing stones, no doubt demarcating a ceremonial space. Similar quartz scattering has also been found at Irish short rows and circles, and in the mountains of Connemara builders erected a number of short rows consisting solely of outsized, standing white quartz blocks (Ruggles, *Astronomy*, 123–24). Lewis-Williams, *Inside*, 215; mediator and "dynamic change," 257–58.

16. "Rayed sun," Herity, *Irish Passage Graves*, 42. Circular and radial motifs are numerous though it's not clear that they all represent the sun. However, some examples can be interpreted with certainty because they are paired with crescent motifs, suggesting sun and moon (Twohig, *The Megalithic Art*, 37). "By making natural phenomena appear liable to social manipulation in the form of cosmological myth, cleverly aligned architecture, and appropriately timed religious ritual, the elite used the predictive value of astronomical knowledge," such as the prediction of eclipses, "as an impressive display and justification of their power and prestige" (Nicholas J. Saunders, quoted by Ruggles, *Astronomy*, 145).

17. "Levers, fulcrums, . . . ropes," Thom quoted by Ruggles, *Astronomy*, 79.

18. "Symbolic function," Ruggles, *Astronomy*, 87 (and see 133, 147, 151, 154). The passage graves at Balnuaran of Clava near the River Nairn, Inverness, Scotland, which make use of white quartz in their construction and mix quartz pebbles with the buried, are aligned with and tilt towards the setting midwinter sun in such a way that it appears to set on top of the tomb itself. According to the archeologist Richard Bradley, "the structure of these tombs establishes a symbolic link between the dead in the burial chambers and the annual cycle of the seasons. By linking the ancestors with the natural order in this way, the people who built the tombs placed them beyond any challenge and linked them to a sequence of events that would continue unbroken into the future" (Bradley, "Architecture, Imagination," 232–33).

19. "Thin thread of sunlight . . . strengthens," Claire O'Kelly, quoted by Hadingham, *Early Man*, 52; Stout, *Newgrange*, 23–24, 29; Burl, *The Stone Circles*, 241 (and see Ruggles, "Interpreting Solstitial Alignments," 12, 17, 19; and Mohen, *The World of Megaliths*, 169). Sunlight penetrates from about a week before to a week after the solstice, but not as fully as on the solstice itself.

20. Etymology, Burl, *The Stone Circles*, 241. Maeshowe shares its midwinter alignment with Newgrange. In the days leading to and following the midwinter solstice, as the sun sets, it sends its last rays directly through the entrance passage of Maeshowe to shine on the rear wall of the central chamber. A distant solitary monolith some nine hundred yards away forms an axis that marks the sun at the solstice as it pierces to the inner chamber of Maeshowe (Tait, "About Maeshowe," online). There is no question, then, of the solar alignment of Newgrange (winter solstice) as well as of Stonehenge (summer and/or winter solstice) (Lewis-Williams, *Inside*, 236).

21. Ruggles, "Interpreting Solstitial Alignments," 12, 19.

22. "Hub about which . . . revolved," R. Castelden quoted by Aveni, *Stairway to the Stars*, 57; "ritualized architecture," North, *Stonehenge*, xxxvi.

23. Burl, *The Stone Circles*, 312–13.

24. Ruggles, *Astronomy*, 44, 37–38, "several degrees . . . south-west," "'corridor' down . . . interior of the monument," 138–39. Such alignment has been confirmed by the recent discovery of two pits that mark the midsummer sunrise and the midsummer solstice, which underscores the intentional positioning of the Heel Stone that aligns with the rising sun on the summer solstice (Ed Caesar, "What Lies Beneath," *Smithsonian* [September 2014], 39–40).

25. North, *Stonehenge*, 519. Though we cannot be certain that these seers were diviners, it is very likely that they operated within some kind of religious belief system. The observations of the sun and moon made from the Heel Stone across Stonehenge could only be made by an individual, which may point to a specialized observer who was part of a religious institution (North, *Stonehenge*, 524). "Astronomical understanding was important," as John North points out, "but it can only have been a part of a larger system" (520), which likely entailed ritual and enhanced the seer's authority. For as Richard Bradley points out, "rituals maintain social divisions by making them part of a timeless natural order" ("Ritual, Time and History," 211). This does not necessarily mean, however, that the sun and moon though objects of fascination were objects of worship as they became in a later age.

26. North, *Stonehenge*, 529; John North says as well, "All things considered, however, it is perhaps rather perverse to try to disengage the Sun as an object from the illumination it provides, and it seems most probable that the latter was seen as significant only because the former was an object of adoration." But the reverse is just as probable and seems to me more likely, given historical precedent and the seer's evident manipulation of rays.

27. Aveni, *Stairway to the Stars*, 90, 182. Evidence discovered by archaeologists at the University of Birmingham in cooperation with the Ludwig Boltzmann Institute for Archaeological Prospection sheds new light on the close relation of the megalith and the sun. Two large pits that probably seated tall stones or posts to mark the sun's rising and setting when viewed from the Heel Stone very likely "defined a processional route" used by the Neolithic peoples at Stonehenge to "celebrate the passage of the sun across the sky at the summer solstice." In addition, this provides insight into "how rituals may have taken place within the Cursus," an immense avenue one hundred meters wide and two and a half kilometers across, just north of Stonehenge, defining a "route of ceremonial processions which took place on the longest day of the year." Clearly, not only Stonehenge but the wider landscape was an area of "ritual focus." We just do not know how the "complex rituals" of light were conducted exactly and "how processions along or around the Cursus were organized at the time Stonehenge was in use" (http://www.birmingham.ac.uk/news/latest/2011/11/25Nov-Discoveries-provide-evidence-of-a-celestial-procession-at-Stonehenge.aspx; accessed 12/6/14). These discoveries counter John North's claim that the "archaeological evidence speaks against the idea of the ritual viewing of midsummer sunrise" (*Stonehenge*, 458). The archaeoastronomer Anthony Aveni's research contradicts this as well (*Stairway to the Stars*, 85), and Richard Bradley maintains that Stonehenge's "role in public ritual" is beyond dispute ("Ritual, Time and History," 211). The evidence for rituals of light and of burial is clear, then, but how the seer braided these together is not.

28. "Its rising rosy . . . Heel Stone," Aveni, *Stairway to the Stars*, 85.

29. North, *Stonehenge*, 528. Archaeoastronomer Anthony Aveni believes that Stonehenge was "built to celebrate the entry of the sun god into the circular sanctuary. It was designed to chart his course as well as that of his rival, the one with the silvered, oft-slivered variable countenance," the moon (*Stairway to the Stars*, 85). Yet there is no evidence of the worship of the sun and moon as gods. In fact, the undeniable alignments of Stonehenge point rather to the celebration of their light not to the bodies themselves. This is supported by the fact that the monument was probably not intended, as Aveni admits, to predict eclipses or to function primarily as an observatory (91). In the age succeeding the Neolithic, the rise of cities generated gods and their priests. In the Babylonian temple of the sun-god at Sippara built by Naram-Sin (2254–2218 BCE), for example, a large disk ornamented with four sets of wave-like rays stands on a small table-altar before the sun seated on a throne. Above is the inscription: "The Disk of the Sun-God and the rays from his eyes." ("The Disk of the Sun-God and the rays from his eyes," Budge, *Babylonian Life and History*, 134).

Chapter Three

The Great Illuminator

The Egyptian Classical Age and the Amarna Period

An Egyptian priest of the Old Kingdom (c. 2600–2150 BCE) sings his "morning litany," recorded on the wall of a king's pyramid burial chamber:

> You awake in peace, Horus of the East [sun], in peace. . . .
> You awake in peace, Harakhti [the horizon], in peace.
> Though you go to rest in the night bark
> you awake in the day bark,
> for you are the one who looks down on the gods:
> there is no god who looks down on you. [1]

And over a thousand years later we hear another song of praise to the sun:

> Splendid you rise, O living Aten [sun], eternal lord!
> You are radiant, beauteous, mighty, . . .
> Your rays light up all faces,
> Your bright hue gives life to hearts,
> When you fill the Two Lands [upper and lower Egypt] with your love. . . .
> You are mother and father of all that you made. [2]

The young Egyptian king Akhenaton himself (1353–1336 BCE) composed this later sun-song engraved on a wall of his courtier Ay's tomb. In a preamble the king tells us that his hymn is addressed to Re-Harakhti, the sun-god, also known as the "Great Illuminator."[3]

LIGHTLAND OF THE SUNFOLK

What experience produced such hymns? How do we account for these singers drunk on sun? Part of the answer lies in the sun-drenched land of Egypt itself, called in many solar hymns "lightland" inhabited by people named on the stela of Amenmose the "sunfolk."[4] I think of Napoleon's quip that "geography explains history." A bit of an exaggeration but his aphorism is a good reminder of something fundamental to our experience. We are in some sense the place in which we live. If we picture the long thin ribbon of fertile land along the Nile laid north to south on a desert where Egyptians have lived from ancient times, we begin to see why these people were sun-obsessed for three thousand years.

Looking out from the Nile, the Egyptians saw the floodplain, reeds and lotus flowers along the river, and further away acacia and date trees, then past the mud-brick houses, desert and light. No mountains intervening, as far as the eye can see; across limestone tableland to the eastern and western horizons, desert.[5] So the central fact of this land is the sun's circuit as it rises on the eastern horizon, passes overhead, and sets on the western, to rise again in the east. Interpreted early on as the birth, maturity, death, and rebirth of the sun, this journey imagined to be by boat ("bark") became central to Egyptian religion and thought. Into the common experience of every ancient Egyptian the sun necessarily intruded, rising over the eastern horizon and flooding the Nile Valley every day with sunlight.

The horizon also, in a way impossible to imagine in the forested mountains and valleys of western Europe or the hills and basins of the British Isles, impinged on the Egyptian consciousness. Its stark straightedge, suggesting permanence and order, dominated their experience. No obstacle impeded their gaze across the desert as it did elsewhere. They knew that in the land of the "wretched Asiatic," for example, "who goes ill with the place where he is," there are many trees and the "ways thereof painful because of the mountains."[6] Ever present, the horizon was the foil to the Egyptian sun's sharp light, an eye-stabbing blaze that arrested the land's people daily. They lived their whole lives between horizons with this eternal presence, truly in a "lightland." And we are exceedingly fortunate that they recorded over the course of three thousand years in their paintings and writings found throughout pyramids and temples this central experience of light.

Especially writings because we no longer have to infer solely from the evidence of paintings and megaliths, as in prehistoric times, the human experience of light. Carved into walls and recorded on papyrus, we find in these documents explicit evidence. We read about the eye of Horus, who is depicted as a falcon, and called the eye of Re (a.k.a. Ra), the sun itself (reminding us of Raven's association with the sun). And his eye, as we would expect, is fierce because the Egyptian sun is fierce. Powerful and relentless, the

Egyptian sun vanquishes all enemies that it sees. Naturally, kings seeking power came to associate themselves with this sun, claiming to be "Re's chosen one," later his son, and eventually the sun-god himself.

But aside from these religious convictions, the Egyptians recorded the physical details of their encounter with the fierce character of the Egyptian sun as well as their interaction with it from a surprisingly naturalistic perspective. Though harsh the sun was also life itself, a fact they could not escape perched on a ribbon of fertile land flanked by desert. Constrained to be partners with its light, fierce and gentle at the same time, they attended closely to its paradoxical character, one resolved in their conception of nature as benevolent.[7] This allowed them to value natural details in themselves, including aspects of the character of light as they impinged on their consciousness.

These details reveal some preoccupations that are familiar from Paleolithic and Neolithic times, such as brightness and rays, which we saw in the king's hymn above. At the same time, Egyptian writings reveal a break with prehistoric concerns. They radically re-value these aspects of light's character. In shifting from the prehistoric to the Egyptian Classical Age and the Amarna Period, we move from the subconscious experience of light's brightness and rays as "common" and "simple" in Reverend Hunter's terms, to the conscious experience of light's character as "glorious" and "mysterious." With the rise of Egyptian civilization, a culture which endured in its essentials for some three thousand years, the human consciousness explicitly reveals its encounters with the glory and mystery of light's character, respectively, recognizing its beauty as a natural phenomenon and charging it with symbolism.

The Egyptian view of the world included both an aesthetic and symbolic or mythic dimension that apprehended light in radically new ways. In the course of a millennium these evolved from an emphasis on the symbolic to the aesthetic, light being imbued with astonishingly personal values. We notice that the older hymn opening this chapter alludes to the sun as all-powerful, reigning over all the gods, journeying in his boat through the sky and the night to return at daybreak in a daily round. And in doing so the priest's litany introduces symbolic figures arrayed in a mythic scheme: Horus as the sun, Harakhty as the horizon. By contrast, the later hymn by Akhenaten not only records an experience of natural light in the natural world as necessary for all growth but makes an explicitly aesthetic valuation, praising it as "splendid" and "beauteous." Later in this chapter we'll see just how radical is Akhenaten's newly observed light. By his time, light had piled on some three millennia of mythological brocade, but he tore it open and exposed the core apprehension of a naked, glorious light.

If one part of the answer to what aspect of ancient Egyptian life produced hymns such as the priest's and Akhenaten's is the sun-drenched land of

Egypt itself, the other is found in the invention of the city-state. The shift in consciousness that brought the aesthetic and symbolic valuation of light's character to the foreground emerged hand in hand with the rise of cites. With the city came urban planning to protect against a fierce sun and the invention of writing, both of which fostered a certain self-consciousness about light.[8] Writing enabled reflection, the ability to step back from one's thoughts, as we've seen in the sun-songs, which is conducive to an aesthetic as well as symbolic response. And the reflection that writing enabled was in Egypt's case reinforced by geography.

Encompassed by the daily cycle of the sun's visitations pouring down its life-giving light and the annual cycle of the Nile's inundations leaving its fertile silt, the Egyptian city-state was blessed by an abundance of food as well as light. As a result, Egyptian civilization beginning with the advent of the city-state (c. 3500–2500 BCE) boasted an amazing sense of stability and continuity of cultural belief and expression over some three thousand years. And this fostered the sophisticated elaboration of an aesthetic and symbolic outlook in reflecting upon themselves a sunfolk.

Their reflections, centered in the sun, were institutionalized as priesthood and kingship. With the city came the creation of a new order in history, the "professional" temple priest. As a consequence, the city was created like the temple as an imitation of the permanent cosmic order. And at the center of this cosmos was the sun-god Re who knew all the cities of the world, their "appearance and incarnations." So playing "jubilant music" every morning, the priests open the gates of the "eastern horizon."[9]

Created hand in hand with the priesthood, kingship merged religion and politics and early on adopted a belief in the sun as supreme power. The king, called the "pharaoh," was himself arch sun-priest. Thus all his actions constituted a religious act, whether conquering enemies or building a road. And the king not only acted on behalf of the sun on earth but "flies to the sky and unites with the sun disk," as the *Life of Sanehut* describes.[10] We hear in this an echo of prehistoric cultures, but where the shaman undertook his journey to gain inner power, a fraught and arduous one not to be undertaken without guides, and the seer materialized his quest, the priest-king appropriates the journey of the sun-god and its power for himself.

Although priest-kings consolidated their religious and political power by such appropriation, of course, they nonetheless testified to a personal experience of sunlight at the religion's origins, a profoundly religious experience of the numinous. They expressed this as being overcome by the beauty of light in its "glory" and gave their experience a symbolic form to ensure the permanence of its "mystery." Such mystery contributed to keeping them in power. The "Eye" of Horus, we remember, is both the bright eye of the falcon to be worshipped as god and the dazzling disk of the sun to be revered for its natural beauty. Furthermore, the eye of Horus becomes the priest-king's eye, establishing an identity with the sun. In the city-state, then, we find ourselves worlds away from the seer who looked on the sun, desiring its power, a one-way gaze. In the new Egyptian Classical Age, the sun-priest looks on the sun who looks upon him, a two-way gaze that establishes a new relationship.

This relationship with the sun finds its full expression in the priest-king's sun-songs. Returning to the hymns separated by a millennium that opened this chapter, we notice that the relationship with light is significantly different between the two. In the first and older hymn, light is bound up with the icon of Horus in the company of gods, though supreme as the "one who looks down on the gods," hence both visible and invisible. By contrast, in the later hymn, actual sunlight appears whose symbol is simply a disk, Aten, alone, with no mention of the gods. The focus of this song is in on the glory of visible light, "radiant, beauteous, mighty," whose "rays light the whole earth." The brightness and rays of natural light are explicitly and consciously named. And the mythic boat-journey disappears, replaced by the two-way gaze. Light is in continuous movement as it participates in the dance of looking and being looked upon.

What accounts for this difference in the hymns? The answer lies in a radical though brief cultural revolution that occurred with the ascension of the young king Akhenaten to the throne one thousand years after the priest of the Old Kingdom intoned the morning litany that opens this chapter. In this revolution, as we'll see, natural light broke free from solar religion, which had prevailed since the first pharaohs, free from the invisible realm of spirit into the visible world of nature. Akhenaten's revision of the traditional solar religion, which the later hymn reflects, though it ultimately did not prevail, was as radical as the Christian revision of Judaism. In its essence, the intense and radiant light personified by the god Re of the old religion was disrupted by the apprehended light symbolized by a simple disk, Aten, of a natural religion called the New Solar Theology. Initiating this, Akhenaten shocked the old order by returning it to the experiential ground of its mythic-symbolic tales, conventional interpretations of a light-experience, from which its traditional religion had grown. In his new apprehension of light, the fascination with brightness and rays persists, but they are re-valued and lead to the discovery of other aspects of its character as well.[11]

STONE-LIGHT ON THE PLAIN, SHADOW-LIGHT IN THE TEMPLE

Looking across the desert, Egyptians from earliest times saw on the horizon the pyramid of Cheops (a.k.a. Khufu). It seems to rise from the edge of the visible world like the sun itself. Dazzling rays ricochet off its white limestone cladding. It is, in fact, a "Luminous Mountain," as Cheops was named. The pyramid's double slopes were angled precisely in accordance with the latitude of the intense Egyptian sun in order to throw maximum light across the desert directly into the observer's eyes. Feeling the force of its rays, the slant surfaces seem to detach themselves from the mountain, slabs of light taking on their own existence.[12]

In this gleaming mountain, light's sheer power and intensity found its fullest embodiment. Egypt, the land of stone and light, united these in the pyramid. Weightless light becomes stone. The soul or essence of light's character, what Egyptians called "Ka," its life force, materializes on the plain. Demanding respect and putting one in a state of awe, the power of Egyptian light risen from the horizon was claimed by the king, who made the name of the sun-god Re part of his name, such as Sahu-Re and Ne-User-Re, and by the Fifth Dynasty incorporated it into his formal title, "the Son of Re." By this he asserted his physical relation to the sun and thus his legitimacy as ruler of Egypt. More than a symbol, the pyramid projected the priest-king's relation to light. As he incarnated light on earth to ensure his body of light in the heavens, so did his pyramid.[13]

Approaching the pyramid, the observer sees its gleaming tip, a four-faceted stone called the "benben," which was echoed by the top of an obelisk at the center of the temple complex.[14] And the pyramid itself is, in effect, a "colossal benben." Its very shape is an image formed by the actual rays of the sun under certain circumstances, such as seen in a dust-particle laden atmosphere (crepuscular rays), and not only expresses the power of light but an intense identification with light. On the road to Saqqara and gazing westwards in the late afternoon, Egyptians had most likely seen this phenomenon, a pyramid of light above the horizon. The pharaoh used the rays of this natural pyramid in the sky, an image of his tomb, to ascend after death into the heavens. "I have trodden those thy rays," the priest-king tells the sun-god Re, recorded in a spell on one of the pyramid walls, "as a ramp under my feet whereon I mount up."[15]

Ordinary Egyptians, once arrived at the luminous mountain, were not allowed beyond the pyramid's gate house because holy ground was beyond the end of a causeway connecting it to the sun-temple in the pyramid complex. Only a priest could pass through its entrance. Leaving this complex that secured the horizon, a purely private space for the priest-king and his court retainers situated on the edge of the visible world, however, we enter the city-center of the world where temple processions welcome the public and their open courtyards the upper strata of Egyptian society. And here is found the "Mature Formal" temple in which the aggressive character of desert light, charging fierce as a lion from a pyramid's face, could be tamed and a new aspect of its character revealed in an interior, the beauty and mystery of mediated light.[16]

Though the priest-king's primary duty, as spelled out in *The King as Sun-Priest*, a manual from the temple of Hatshepsut, was to "satisfy the gods" and "administer justice," the temple first had to entice the sun-god to descend from the sky and inhabit his statue that stands inside its sanctuary. Hence temples were generally aligned with the axis of the sun in Egypt.[17] Only as the seduced god entered did the temple become, as intended, his house. Only then was his power concentrated and available for effecting order and justice in the kingdom. Thus the temple had to function as a god-magnet. It was designed so that when the sun in the "guise of Horus," the hawk, "sails through the heavens" and beholds the work of architecture, as a contemporary priest expressed it, the sun-hawk will pause "before this temple" and his heart be "filled with joy as he looks at it. Then he becomes one with his statue." This is the fulfillment of daylight dwelling in the center of the city forever.[18]

Light, upon entering the sacred precinct of the temple, materializes as a god—the cult statue standing at the heart of the temple. But love of beauty precipitates his act. It is not prayers or supplications, significantly, that bring light down into the temple but its beauty. The Egyptian writer, describing this seduction of the god, projects onto the temple his desire for its power to capture and materialize the light that sails in the remote space of the heavens and which he loves.

God-magnet and light-attractor, the temple's symbolism and physical beauty are inseparable. The house of the god incarnates the "celestial horizon from which he emerged to bring light into the world." That is, the temple brings the distant horizon of rising and setting sun inside the temple itself. And this light that first brought the world into being, which the temple attracts, is the same light appearing every morning.[19] From "quartzite and black stone" walls to gleaming calcite floors inside, the temple modulates light's body of rays and brightness for a most specific purpose—beauty. Describing their temples, the Egyptians used many words that can be translated as "beautiful," "magnificent," "sumptuous," and other synonyms for "beauty."[20]

With these words of witness to the beauty of light, explicit aesthetic emotion enters history for the first time. The Egyptian priest in a reflective moment away from the pressure of his day-to-day duties would undoubtedly be taken by such beauty. For it is an expression of an encounter with the character of light that he no doubt shared with others of his class throughout the three millennia of Egyptian culture. The qualities he apprehended in light were those he attributed to the sun-god in the sun-temple. Thus its exterior was dedicated to the seduction of light, its interior to the materialization of light, a being of beauty, as was the sculpted Osiris lying face up in the temple of Denderath at the bottom of a light shaft.

Entering the interior of a temple, the priest experienced a new aspect of light as well, its mystery. Superb New Kingdom (1539–1075 BCE) architects set out to build one of the most glorious light-attractors ever and succeeded in the Great Temple at Karnak dedicated to the sun-god Amun. Scene of a daily ritual conducted by the priest as well as monthly and annual processions witnessed by the elite "sun people" of the city, this temple continued as a religious and economic center throughout the period.[21]

After the priest purified himself in the "house of the morning," he passed through the mammoth gate of the temple, called a "pylon" formed by two massive, trapezoidal blocks of stone rising on each side of a tall door and towering above it. As he passed through this gate he became, in a sense, the sun itself. For the pylons form a giant hieroglyph, the word *akhet* meaning "the horizon." Proceeding west to east on its central axis, he found himself in a large sunny court and at its far end ascended a ramp into a hall populated with 134 massive pillars, called the "hypostyle hall."[22]

Once inside he experienced a subtle drama. Throughout the interior leading to the sun-god himself, light's character as a living power of complex beauty is revealed in a light-shadow dance. A power intensified by an atmosphere of mystery as this dance continues throughout. By means of "sunk relief" all around him (obverse of raised bas-relief) chiseled on walls, columns, and obelisks, this dance embraced him. The sunk-relief sculptor did not gouge the outline of an image out of the stone, leaving a shell-like trough, but cut it out at a ninety-degree angle. This left a sharp edge on one shoulder of the hollow that catches the light, producing a line bright as neon lighting. The other shoulder leaning into the image formed a line of pure shadow. The contrast is stunningly dramatic.

The sunk relief like the temple itself is not only a "light-catcher," then, but a light-caster as well, extremely sensitive to changes throughout the day in light and shadow. As day progressed, the shifting light in the temple partnered with mysterious shifting shadows. In their dance they moved in and out of niches, exchanging places, alternately erasing and facing each other.[23] The body of light includes shadow, which is not the same as darkness. The character of light encompasses shadow, revealing a mysterious aspect of light otherwise lost, whereas darkness banishes light.

The Egyptians expressed this new relation to light in the technique of sunk relief, which they invented. In Neolithic experience as embodied in their monuments the direction of light claimed prominence, as they channeled it down precise paths to touch on a sacred stone. Egyptians emphasized instead the interplay of light-shadow, largely unconcerned within the interior spaces of their temples with the direction of light.[24] Once light has been seduced and enters the temple, it reveals its mysterious body, as the priest passes through the hypostyle hall, its sunk reliefs rhythmically catching and casting the light. More than rays and brightness, it is a numinous light-shadow presence having great emotional resonance, especially given the importance of shadow in Egyptian culture. In short, the Egyptian temple architects achieved expression of a new dimension of light's character, creating a correct balance of the beautiful "light of life and the shadow of mystery."[25]

Big changes, however, were to visit the classical temple. Finishing work on the gate to Karnak named for his father, the new king Amenophis IV had artisans depict him as the traditional pharaoh with the falcon-headed sun-god, but directed them to carve in bold columns a new epithet for the god: "he who rejoices in the horizon in his name 'Sunlight that is in the Disc.'" Sunlight named. Light embodied simply in a "Disc"?[26]

Not the intense stone-light ricocheting off the face of pyramids or the modulated light-shadow dance in the traditional temples, but something different, a naked natural light whose character is not bound up with pyramids and temples. A central aspect of light's character, this had been obscured by monuments to its power and by the priest-kings' determination to appropriate this. But now a new relationship with light counters the symbolic. Egyptians responded to natural light directly, as we'll see, celebrating its beauty not as ferocious or mysterious but creative, not as light to be modulated but to be celebrated in itself. And no god was necessary to this celebration. Hence a return to Egyptian religion's originating experience of naked light threatens the old order.

NAKED-LIGHT OF THE NATURAL WORLD

Everything about the new king's physical appearance seemed to be exaggerated. His skull was elongated, his lips fleshy, ear lobes lengthened. And his eyes slanted so much they gave him the appearance of being from another world. It's no wonder that his father Amenophis III forbade his image to be carved on any monument. Yet this strange young man, Amenophis IV, ascended to the throne in 1377 BCE. Perhaps it was to be expected, then, that he initiated a revisionist religion and, changing his name to Akhenaten, "Glorified Spirit of the Sun-disc," moved to a city he had built at Amarna. Dedicated to the sun, he named it Akhetaten, "Horizon of the Disc."[27] And here he built a new temple, "House of the Sun-disc," unlike anything the traditional priests had ever seen, which celebrated an aspect of light's character that had been buried for millennia in mythological tales of the sun-god.

Called the Temple of Aten (whose emblem was the sun-disc) where the High Priest of Akhetaten held the same title as the High Priest of Heliopolis—"He who Sees the Great One [sun]"—this sun temple departed from the traditional layout. Instead, its center was an extensive sun court open to the light, adapted from the Heliopolis temple, and the sanctuary contained not a statue but the gleaming sun disk alone, emblem of the solar disk, called the

"Aten." In the center of the courtyard large enough to accommodate over four football fields was a raised platform on which cult offerings were made to the sun-god, establishing contact not with a statue but directly with the sun seen by all in the sky. The naked sun in all its glory thus displaced the statue, the traditional cultic image.[28]

Instead of the classical temple hosting the uncanny dance of light-shadow we've seen, Akhenaten's Temple was completely open within its high walls to the sun. Every ritual was bathed in radiance. Rather than being dedicated to mediating the rays of the sun, however, this temple was devoted to welcoming them. In a radical break with the past, Akhenaten, as soon as he rose in the morning, filled the temple with himself "by means of his rays," as one of his Akhetaten boundary stelae states. He dispensed with mystery and celebrated the experience of pure beauty. For the celebrants observing the ritual as well as worshippers in the courtyard, however, this was not necessarily a pleasant occasion. A contemporary Assyrian king, Assuruballit, complains in a letter to Akhenaten that his ambassadors were kept standing for the liturgy in the blazing light, god or not, so long they nearly died "under the sun."[29]

Clearly, the young and little known Akhanetan was a king of rare individuality, perhaps because he had been so long confined to the court and ignored by his father. The notion of the king as son of the sun-god Re had prevailed since the Old Kingdom, but the new king identified with the sun itself, becoming its "Spirit." And though his father, who called himself the "Dazzling Sun," had claimed to be the sun-god before him, the intensity of Akhenaten's identification and insistence on the exclusive nature of his relationship was new. He reveled in the beauty of light, being "fairly ecstatic" in fact. This complete identification with sunlight, exalting his personal experience of light out of which he created his own religion, elaborating the "role of the sun as light," marks a momentous turn to the personal. A swerve from tradition, as we'll see, that precipitated a revolution in the human experience of light.[30]

Moved by the "pure presence" of light, Akhenaten insists on a directly observed natural light. Seen first, and only then made the fundamental principle of cosmic and earthly life, this is evoked by the Aten, depicted as a sun disk emitting rays (see Figure 3.1). In short, Akhenaten in about 1370 BCE, having moved to the desert area of Amarna where he built his own city and new temples, revolted against the composite god Amun-Re and the political power of his priests back in Thebes/Karnak as well as the spiritual power of Heliopolis home to an ancient sun cult. And this cleared the way for an even bolder move. He elevated the sun-god Re as the *only* god, light itself, whose icon was the sun-disk called the Aten. As light comes forward, therefore, revealing itself in the rays of the Aten, the god recedes.[31]

This act of reducing the sun-god to an inanimate material disk, in effect stripping the god of his mythological robes, was pivotal. In so doing Akhenaten returned to and clarified the experiential basis of the long tradition of Egyptian sun worship. Restoring the individual experience of sunlight, bringing it onto center stage and celebrating its beauty, could not but shock the traditional priests. Who could imagine a world with no gods but the sun, whose icon was simply a bright disk, no bathing and clothing the mythic god in his own house? Thus as Akhenaten testifies, all the gods "have ceased one

Figure 3.1. Akhenaten and Nefertiti worshipping Aten. c. 1535 BCE (Egyptian Museum, Cairo). ***Jürgen Liepe, Berlin***

after the other." Thus he can replace their statues in the temple sanctuary with the sun-disk and address the sun itself in triumph, "Hail to thee in thy radiance." Sunlight now revealed in all its splendid nakedness.[32]

Akhenaten insists on the "aloneness" of Aten. The sun-disk is not only unique but alone, not bound up with the company of gods. Isolating the radiance of Aten, brightness and rays now weigh in the human experience of light as never before. Naked, they do not so much refer to the numinous as now take possession of their natural character, which is beauty. This does not mean that the Aten, however, lacks a religious dimension. It remains, after all, the only "god." But a new kind of god, radically different than the traditional gods in that Aten is the god not of a mythic creation in the past but of the cosmic cycle of creation in the present as well, which we can observe. "Aten" is the name not so much of a god as of the "solar orb" and is thus fundamentally perceived as "light radiating from the sun." Aten literally brings the earth to life daily, that is, sustaining by his radiance its *natural* processes. Plants and animals as well as the sunfolk themselves owe their life to this "god" of natural creation, namely sunlight itself.[33]

Religious truths, that is, make way for natural facts. Akhenaten's was a revolt on behalf of visible nature, specifically sunlight. And he was driven by his relationship with this character, a profound light-experience. Having completed his necessary preliminary project of negation, his stripping away of mythic trappings, he launched one of affirmation. This not only moved sunlight onto center stage but elicited an individual response to which Akhenaten gives expression.

The character he encountered was new in the Egyptian experience of light, which had since the Old Kingdom been largely informed by religious symbolism, however intensely powerful in the desert and hauntingly dramatic in the old temples. In the course of Akhenaten's brief reign, we witness a relationship with light pervaded by an aesthetic sensibility that subordinated the mythic. The individual's apprehension of visible light as a beautiful phenomenon in the world, capable of evoking an "ecstatic" response, becomes the starting point. Here Akhenaten returns us to the experiential basis of Egyptian sun worship, a personal experience of the "radiance" of sunlight, a beauty that evokes feelings of love in the beholder. The Amarna hymns to the sun use not only the word "beauty" but "love" as synonyms for "rays," emphasizing the personal.[34]

Sunlight appears to Akhenaten as a self-created "god" as well as a god who sustains the creative natural processes in our world, making possible a personal relationship with radiance. Because light connects creator and creation, an intimate human relationship exists with sunlight. Significantly, this is a god who manifests himself solely in natural phenomena. Hymns exclaiming, "You appear beautiful / in the eastern part of the sky . . . / sparkling," clearly give expression to an encounter with the profound beauty of an observed, natural aspect of the sun's light. We witness here an emotional as well as intellectual response, involvement, and detachment. And the new use of verbs in Egyptian to express visible, phenomenal qualities, such as "sparkling," brings the visible to the forefront, upsetting the traditional religious commitment to balancing visible and invisible qualities.[35]

Naked sunlight thus displaying its "natural" rather than religious character. For Akhenaten sunlight is both more and less than the god's beneficent gaze on the world. Whereas earlier theological interpretations dominated the natural phenomenon, the emphasis is now reversed. The Aten is "the lord of sunbeams, maker of brightness," according to one inscription on the disk. Icon of this new emphasis, Aten invites observation of the effect of sunlight. And this leads to apprehension of the nature of sunlight, the one who "causes everyone" to live, and people are "never sated with seeing him." Radiance is intimately related to us as well as being a phenomenon in itself. In art of the period, rays are depicted ending in hands, a striking image of Akhenaten's experience of natural light as both beauty (physical) and love (religious).[36]

Underlying his new experience of light is the certainty that a single "natural" principle inherent in the Aten's "aloneness," light's independence from the world and the gods, not only informs the multitude of natural phenomena around us but governs all existence. Light creates time and space as well as sustains life itself, plants and people. The sun *causes* abundance of life and does not simply bless the abundance of his creation.

Unlike any god, light is embedded in visible nature. Yet like a god, it is separate from nature. Akhenaten's experience of this paradox finds expression in the image of the Aten. Light, "The Living one," appears in the sunrise in his "identity as the light that comes in the sundisk."[37] Thus Akhenaten initiated not so much a religion as a natural philosophy, though for him the one did not wholly preclude the other. As we'll see, the imagery of his "Great Hymn to the Aten" reveals a commitment to both. His hymn to the sun-disk celebrates his relationship with a fundamentally creative and kinetic character. As a result, his solar hymn cuts against the grain of tradition and reveals the very inner character of Egyptian light.

AKHENATEN'S "GREAT HYMN TO THE ATEN"

Hymns to the sun in Egypt preceded Akhenaten's, of course, such as the one we heard opening this chapter. They had their origins in the "morning litanies" of the Old Kingdom, which focused on the sun-boat's journey from its rising to its setting and rising again. For Egyptian priests experienced the sun in its relation to the divine forces of nature. And the hymns themselves, which constituted sacred knowledge expressed in formal language, were canonized over time and changed very little in two thousand years owing to the continuity of Egyptian culture. By reciting daily the solar liturgy the priest participated in the gods' journey and by that kept it going, which was the same as preserving cosmic and social order as well as the life of the king and his subjects.[38]

These liturgical hymns to the rising and setting sun continued to be copied and recited during the "golden age" of solar hymns (c. 1500–1200 BCE), which includes the Amarna Period (1353–1292) and Akhenaten's reign. But parallel to them a new kind of hymn to the sun appeared. Exalting light accessible to all rather than to the priest exclusively with his esoteric knowledge of the sun-boat's journey and its symbols, this hymn emphasized within the context of natural processes the sun's light and motion. One sung by the Egyptian architect Suti serving under Akhenaten's father, for example, engraved on a granite stela outside his tomb, reveals not only sunlight itself but a personal response to it, one resisting the pressure of the symbolic tradition and allowing direct observation to surface.

Suti addresses the sun dawn directly: "Your brightness opens the eyes of the herds." Subordinating the symbolic to the physical, an implicit aesthetic response informs Suti's religious experience. He alludes to the sun-god's traditional journey but in a personal way, making us aware of the actual physical sunlight on the body. Its "rays" on his face are somewhat mysterious, "unknown," as they are in the temple, but in their "brightness" unmistakably beautiful. The sun is let out of the boat and is now available to direct aesthetic experience.[39]

From this traditional yet freshly plowed soil, by men such as Suti, blooms Akhenaten's great celebratory and revelatory hymn attesting a radically new experience of light. We can see how revolutionary this is when we compare it with another well-known hymn, Psalm 104. Just a couple of hundred years after Akhenaten, the Hebrew people to the north, ironically, appropriate his hymn to express their conviction of a light inseparable from divinity. They insist on the mythic expression of light's character, reversing Akhenaten's emphasis, as they exalt the invisible over the visible. Consequently, the Psalm drains the celebratory element from the Egyptian hymn, which opens with, "O thou great living sun disk . . . thou risest beautiful . . . shining on the eastern horizon, having filled every land with thy beauty!" And Akhenaten being moved aesthetically says, "Thou hast made the heavens afar off in order to shine in it."

The Psalmist parallels this language but makes a significant change, "O Yahweh my god! Thou art very great," being moved religiously, who "coverest thyself with light as with a garment." Light is its own presence in the former, "beautiful," "shining," "filling the land," while in the latter light is simply Yahweh's "garment." Akhenaten's hymn, ambiguous about the "sun disk," celebrates light; the Hebrew Psalm, unambiguous about Yahweh, worships the "god" who is ultimately invisible. For Akhenaten the reverse is true, the only knowledge of god derives from the visible.[40]

Inscribed in three columns of hieroglyphs on the doorway to a tomb, we have by a stroke of luck Akhenaten's very words. And in his revolutionary hymn, though adapted for a courtier's recitation, we have for the first time in history an individual's own testimony of his experience of light. This hymn dramatizes his personal swerve from tradition. The glimpses we've seen of Akhenaten's response to light make clear his move from the mythical to the physical, breaking radically with the past, which he consolidated in his move to Amarna.

He accomplishes this by taking, first, the rational path we saw earlier, stripping symbol and myth from his childhood religion in order to expose naked light. He insists on the visible. Thus in his relationship with light he supplants the mythic with a contemplative view of nature. And by this he experiences light in its essence as self-created (as were the Egyptian gods from earliest times), the Aten having "built himself with his own hands" as the hymn says, and created the world with his sheer presence. This prepared the way, second, for Akhenaten's personal path, apprehending nature directly. Fundamentally visible, light is not a symbol in a religion or an icon in a mythology for him but a creative and kinetic presence in the world. Thus the sun-god Re is not the subject of his "Great Hymn." Rather, *sunlight* named Re "shining in your form of living Aten," as Akhenaten acknowledges, plays a lead role in the drama of nature's cycle of sunrise and sunset. Unlike the priest's, Akhenaten's is one world, not two.[41]

Akhenaten's new insistence on looking at light, rather than through it to a god, brings his "observing eye" on stage. Traditional cosmology had room for the eye as symbol of the sun shining, "looking" on us, the arresting eye painted on many Egyptian coffins. But it did not have room for the reverse, the physical human eye observing sunlight. Just as he shifts the emphasis of the traditional solar hymn from god to nature, so he shifts from the "eye" in heaven to the literal eye of the human observer on earth. If reality is reduced to the visible, as Akhenaten does in his hymn, then he must include the eye responding to the light. And he is a player in the drama of nature from the first line of the hymn: "Splendid you rise in heaven's lightland, / O living Aten." Akhenaten praises him as "great" and "radiant" in our observed world, "Lightland," in a magnificent celebration of life. We see earth "brighten" at dawn and the sheer "shine" of daytime reveals beasts browsing, birds flying from their nests, and fish darting in the river. "All eyes observe," Akhenaten reminds us more than once; "All eyes are on [Aten's] beauty." In lightland his eye collaborates with the eye of the sun. To exist is to exult in beauty, giving in to the aesthetic response, and celebrate the activity of light.[42]

The personal tone of Akhenaten's hymn reveals an intimate relationship with light. He conveys a sense of wonder at the beauty of observed light as well as a palpable ecstasy. He is moved by the world's utter dependence on light, deriving his philosophy from this "central principle."[43] Akhenaten, that is, comes into a revelatory apprehension of its brightness and shine, rays and radiance. Whatever he discovers of the nature of light's character and effect on him originates in this encounter, a profoundly aesthetic experience that for him was no doubt inseparable from an intimately spiritual one. And his new apprehension of its character reveals two aspects that almost overwhelm Akhenaten, its boundless creativity and kinetic power.

Carried on "rays," light's creative force in Akhenaten's experience derives from its radiance. Potent enough to create time and space, it also maintains intimacy. Sunlight "feeds the son in his mother's womb" and "soothes him to still his tears." Time created in human terms. Here is the power we've seen in the stone-light of the desert, as the rays create life itself. Yet not simply power, this is a generative potency, a new aspect of light's character that includes tenderness as it "feeds" and "soothes" the child in the womb. And we get a hint a few lines later of the mysterious modulation of the temple-light we've seen, as Akhenaten acknowledges that though visible, some of the rays' many deeds are "hidden from sight." But a gentle note in his voice has colored this aspect as well, having just shown us rays sensitive to the time "When the chick in the egg speaks in the shell," emerging "To announce his completion." This is not only moving but unprecedented. In such a closely observed event, the chick "walking on his legs" from the shell, the marriage of light and time in Akhenaten's experience of light is stunning.

Grounded in his sensuous view of the world and reality itself, Akhenaten also shows us the marriage of light and space. Again, the creative power of light is on display, as its rays make the "far sky to shine" and further, make of themselves "millions of forms"—the "Towns, villages, fields," and even "the river's course." But as with time, Akhenaten precedes this astonishing observation of light's character with a touching image, saying that the sun's rays "nurse" the fields. Potent and gentle, this light whose "beauty" attracts "all eyes" thrives above all on its intimate relationship with the sunfolk.[44]

Just as the rays look on the world of time and space they have made, the chick completed and the town and river formed, so its creatures look on the rays. Uniting all existence in the common space of this reciprocal relationship of seeing and being seen, light is the great intermediary. This "communicative" nature of light is for Akhenaten an essential aspect of its creative character.[45] What matters is not the Aten's religious meaning, but how light behaves in the world. All depends on mutual responsiveness. Akhenaten doesn't concern himself with a priestly interpretation of light's involvement with the origins of the world in some primordial time but with light's actions in the here and now. He observes how the world comes to be by light, not *came* to be in the past. We're caught up with him in the drama of light's daily acts of creation rather than reminded of a single, primeval creative act. "In the beginning" and "in the end," their speculated netherworld yawning open below us in death, are irrelevant to this light that walks gently and carries its immense power gracefully without judgment throughout the world.

Gentle creative power, certainly, but Akhenaten doesn't let us forget light's immense kinetic power. The light "Risen, radiant, distant, near" is a dervish. More than brightness, his is motion incarnate. Though starting from a contemplative observation of light, Akhenaten, as he probes its character further, seems overwhelmed by this relentless force. The "rays embrace the lands" for starters. Then Aten "dispels" the dark and "rouses" the people of the Two Lands (Upper and Lower Egypt) and everyone "sets out to work." Meanwhile, he continues his own work. He "opens" the roads and probes the sea, "makes seed grow in women" and "supplies" everyone's needs. Not satisfied, "He makes waves on the mountains like the sea, / To drench their fields and their towns," makes the seasons, and ultimately "millions of forms." And he does every day.

The visible world is, in fact, fundamentally a "becoming" world. Light-land is not only generated but driven by the cosmic motion of light. All that we see is a function of light. This means for Akhenaten that god and the world are bound up together in the energy that is the Aten and the world together. They are not mirror images of each other, that is, as are the divine and social orders of traditional Egyptian religion. Nor are they radically separate from one another as Hebrew monotheism insists. Rather the Aten and the world are inextricably bound up together. So important to Akhenaten is this principle of reality that he makes it explicit. The name he attaches to the morning sun at one point, *kheperu*, means "to become, to come into being, to develop," a natural process not a finished product. Thus we see all come into being by the kinetic force of the Aten's rays, the child and the chick as well as village and field and river.[46]

The personal voice that infuses Akhenaten's hymn exploring his encounter with light in the act of becoming, the overwhelming kinetic energy of sunlight startles us at first. After all, he is the all-powerful Pharaoh, the king of the land. Speaking in everyday language, unlike the priests', and noting every mundane particular of the earth, from birds and herds to differences in people's skin color and speech, which elicit empathy, he seems peculiarly involved with everyday life. We can't help but wonder, especially because he departs radically from the traditional solar hymn, why this is. All becomes clear, however, when at the end of his hymn, addressing the Aten, he bursts out: "You are in my heart, / There is no other who knows you, / Only your son" Akhenaten "whom you have taught your ways and your might."[47]

Thus he identifies completely with his discovered light. This accounts for the intimacy of his hymn. Knowing essential aspects of light's character, which no one else on earth does, he sees himself as this light. Others observe the light observing them, but only he knows light's true character. This knowledge lies outside his hymn's drama of the daily cycle of generation. The world may be in flux, by light becoming world, yet light itself is permanent. And he appropriates this for himself, taking the Aten into his "heart." Akhenaten is the still point in the daily cycle of the coming and going in the world of "divine energy." With this comes one scholar's radical claim for Akhenaten, "There is no natural principle but light, and no god but Akhenaten." Unlike his subjects, as the sun-god's son he is a completed creation and herein lies his authority as king.[48]

His personal relationship with light is fundamentally sensual. Akhenaten contemplates light's beauty *before* he commits himself to it as the first principle of reality. "God is nothing else but the sun," for this young king that is, and "he is also nature," as the great Egyptologist Jan Assmann concludes.[49] In this sense Akhenaten, who liberated himself from the authority of Heliopolis and the priesthood of the solar cult, was the first in history to find a path out of religion itself. He was able to do this because his intimate relation with new aspects of light's character, once stripped of its mythological brocade, denied the world the theological significance that the old religion accorded it. Akhenaten's communing with physical light in a physical world, that is with the Aten, exposed the essential experiential ground of sun worship while gutting the cult of the sun-god.

INNOCENCE LOST, LIGHT AND DARKNESS

Akhenaten's revelation of light's character remained largely personal, however, and failed to bear much fruit in his own country. At the end of his reign and those of his grandson Tutankhamen and great grandsons during the remainder of the Amarna Period, the pharaohs who came to the throne lost no time in reinstating the priesthood with its traditional power and perks and their god Amun-Re. He was pulled naked out of nature, his religious robes restored, and re-integrated into the society of gods, albeit as supreme.

The sun becomes once again "chief of the gods," donning a god's apparel and taking a god's journey. The essential human experience of light out of which Egyptian sun worship had arisen being once again lost sight of. And light the Creator is no longer married to creation. This is, of course, antithetical to Akhenaten's experience of light as immanent in creation, which rendered heaven and the underworld unimportant to him.[50] So as Akhenaten's contemplative, intimate relation with a fundamentally creative and uniquely kinetic light gives way to the mythic, his natural philosophy disappears under the flood waters of religion.

And with this return to ancient tradition, darkness takes on importance. Akhenaten's conviction of the utter dependence of the world on a physical light had diminished the importance of darkness. So for him there was no longer a hereafter or realm of the dead because these worlds had no independent existence. They were continuous with that of the living, the Aten shining in all.[51] But as this conviction weakened and traditional mythology regained its former strength, darkness came to dominate the psyche. In native Egyptian experience, the complementary relationship of light and darkness followed closely natural processes, defined by the daily circuit of the sun. And as we've seen, Akhenaten had taken this tradition back to its experiential origins, each phase of this being simply an aspect of the natural cycle, involving no divine intervention, and thus innocent. Given that the alternation of light and darkness was simply a natural fact, darkness like shadow carried little ethical charge. The modulated light of the formal temple with its shadow dance generated an atmosphere of mystery not fear.

But this was destined to change after Akhenaten's death as a consequence of the revived Egyptian mythology. An innocent darkness, which for him was no more than absence of light, became the adversary of pyramid and temple light. And where originally no ethical values had been assigned to light and darkness, now the sun came to be associated with justice and the good, whereas darkness was identified with their opposites. Woven into Egyptian religion from earliest times was a strain of light and darkness as physical beings. And these beings like Horus, the solar deity of the Early Dynastic kings, and Set, god of darkness, had engaged in combat.[52] Such a relationship invites symbolic and mythological associations with life and death, which the later priesthood took a step farther. Abstracted from the natural cycle—day, night, day—these associations harden. The emotional meaning of light condenses to a fixed symbol that demands an equal but opposite symbol, a dreaded darkness. In turn, fear spawned a theology of light and darkness, introducing a dualistic relationship.[53]

Marking the severance of the light-dark relationship from direct experience, such a conventional dualism of light and darkness shaped much of the subsequent imagination of light. Interpreted as good and evil, the conflict between light and dark came to grip the Persian Empire. Passed on by Iranian to Indian mythology, it influences to this day Mithraic and Christian mythology. It is also reflected in Hollywood, where in *The Empire Strikes Back* (1980) of the *Star Wars* trilogy, the battle between light and darkness, life versus death, good versus evil, was once again waged. A radiant Luke Skywalker wielding his light-sword fights a black-robed and black-hooded Darth Vader.[54]

Emergence of a light versus darkness dualism in Persia in about the sixth-century BCE, then, reflects a shift in human experience. They are no longer complementary in this later period, which we could call the "Age of Sages."

A change largely owing to individuals across the world, such as Isaiah in Israel, Plato in Greece, Siddhartha in India, and Confucius in China, who formulated many of the religious certainties and philosophical questions still with us.[55] The sages extend the self-consciously aesthetic and symbolic response to light inaugurated, respectively, by Akhenaten and the Egyptian priesthood. But while the aesthetic stance had returned the boy-king to nature, the restoration of traditional Egyptian religious symbolism pointed his successors away from nature.

Akhenaten's experience of light led him to discover an essential aspect of its character, which we could call "self-awareness." An attribute embedded in Aten's didactic name, meaning "the living One," who "rejoices on the horizon, in his name (identity) which is Illumination" from the "solar orb." Light rejoices in its own beauty, which is central to Akhenaten's imagination of light. Thus Aten was not the disk itself but as we know, natural "light radiating from the sun."[56] Self-created light, that is, reveled in its own radiant powers of creation, sheer naked energy revealed. Akhenaten's encounter with this self-created light was a new experience in the world, one that classical Greece and Rome give full expression to, as we'll see in the next two chapters. And one our own time is enamored with. The light-artist James Turrell, prominent in the California "light and space" art movement (debuted 1971), for example, paid homage to Akhenaten by entitling the centerpiece of his 2013 Guggenheim Museum exhibition, "Aten Reign."

Akhenaten's natural philosophy, born of aesthetic wonder, did not preclude his total identification with the Aten in which he declares himself "god." But it did exclude the religion rooted in the Egyptian culture that resumed after his death and that dominated the religious symbolism of the later age in which this unity of man and god-light splits. As a result, natural philosophy and religion take forking paths in the Age of the Sages. One follows the way of understanding the universe, which Greek philosophers like Thales took, though not directly influenced by Akhenaten; the other takes the way of relating rightly to the universe, which religions initiated in this age followed—Judaism, Buddhism, Hinduism.[57]

The latter proved to be, of course, the path most traveled. In the next millennium religious and philosophical symbolism becomes the primary expression of the human response to light. Conversely, light became a virtually universal symbol of religious experience, whether institutional or mystical. Thus the short-lived Amarna tradition of Akhenaten, rooted in natural philosophy, gives way to the mythic and fades from view not to be seen again with clarity, though we're given glimpses, until the Renaissance. The irony strikes us that the lonely, homely boy king who discovered a way out of religion in natural philosophy during his brief reign is followed by an age dominated by religious forms.

NOTES

1. Plas, "Hymns," *Oxford Encyclopedia of Ancient Egypt.*
2. Lichtheim, Ancient Egyptian Literature, II, 91–92.
3. Assmann, *Egyptian Solar Religion*, 112.
4. "Lightland," "Naktamum Book of the Dead," Assmann, *Egyptian Solar Religion*, 45; "sunfolk," "The Great Hymn to Osirus," Lichtheim, Ancient Egyptian Literature, II, 84.
5. Michalowski, *Art of Ancient Egypt*, 43; Parcak, "The Physical Context of Ancient Egypt," 11.
6. "Wretched Asiatic, . . . because of the mountains," Wilson, *The Culture of Ancient Egypt*, 112.
7. Gazing east and west from their Nile River Valley home unthreatened by mountain crags or ocean abysses, they did not share the anxiety about nature of their contemporaries, the Sumerians and Akkadians of Mesopotamia, who occupied the less benign landscape of the Euphrates river valley (Smith, *The Art and Architecture of Ancient Egypt*, 4).
8. Planning a city, such as the Sumerian city of Ur, which came of age during the time of Egypt's Old Kingdom in the third millennium, necessitated viewing light in a new way. This required a combination of observation and analysis, which assumes a certain self-consciousness. Like Egypt, the Mesopotamian culture shared the problem of protecting against a fierce sun. Accordingly, they laid their streets out in a diagonal pattern, thereby creating a "shadow network." This made socializing and working in the city comfortable compared to the countryside (Shepperson, "Planning for the Sun," 364, 367–68).
9. "Professional," Campbell, *The Masks of God: Primitive Mythology*, 146–47; "appearance and incarnations . . . eastern horizon," Assmann, *Egyptian Solar Religion*, 19.
10. "Pharaoh," Quirke, *The Cult of Ra*, 82; "flies . . . unites with the sun disk," Quirke, *The Cult of Ra*, 17, 20.
11. Solar religion was an essential element of Egyptian culture from the first Pharaohs and their pyramids and traditional temples of the Early Dynastic Period (2950–2600 BCE) through the reign of Akhenaten's grandson Tutankhamun (1332–1322) and the last Dynasty (664–525) until its demise. But a remarkable revision occurred in the New Kingdom (1539–c. 1075) during the Amarna Period (1332–1322), wrought by Akhenaten. I follow Shafer's chronology, *Religion in Ancient Egypt*; "New Solar Theology" is Assmann's term, *Egyptian Solar Religion*.
12. "Luminous Mountain," Levin, "The Eye of Ra," 18; Giedion, *The Eternal Present*, 505.
13. Wilson, *The Culture of Ancient Egypt*, 88.
14. Quirke, *The Cult of Ra*, 115–17, 129; Manuelian, *The Ancient Egyptian Pyramid Texts*, 431. Preserved as the top of an obelisk, its form was a mini-pyramid made of quartzite sheathed with beaten electrum, an alloy of silver and gold, and remained part of the temple complex throughout Egypt's long history. Luminous minerals, such as quartzite, were believed to be "solidified celestial light," an inheritance from prehistoric times, and they were used for cult statues of the gods as well as of kings and queens like Ankhnesmerira sculpted in translucent calcite. Representing a ray of sun, the obelisk stood wherever the sun was worshipped, such as at Heliopolis, the sun temple of Abu Gurab, and later at Karnak and Luxor (Levin, "The Eye of Ra," 15).We know from a typical solar obelisk inscription that the "benben" was "activated" as the potent presence of the "primeval mound," the original dry earth of creation, when the first rays of the sun caught it and the polished surface flared. For this reason, the pharaohs would go to great lengths in building their pyramid tombs to ensure that sun temples were in close proximity. If this was not possible, they secured the sightlines to them. So potent was their identification with light as presence and power that they were compelled to bind together in this manner their burial place and the home of the sun-god on earth (Quirke, *The Cult of Ra*, 27, 128, 138).
15. "Colossal benben," Fakhry, *The Pyramids*, 8; "remain for the span of eternal sameness," Allen, "The Cosmology of the Pyramid Texts," 26; "I have trodden . . . eye of Re," Levin, "The Eye of Ra," 16.
16. "Mature Formal," Kemp's term, *Anatomy of a Civilization*, 248.

17. Shepperson, "Planning for the Sun," 371, 373, 375.

18. "Guise . . . through the heavens . . . becomes one with his statue," quoted by De Cenival, *Living Architecture*, 86.

19. Spencer, "Priests and Temples: Pharaonic," 267; "celestial horizon . . . light into the world," De Cenival, *Living Architecture*, 86; Levin, "The Eye of Ra," 12.

20. Wilson, "Temple Architecture and Decorative systems," 781; "beauty," De Cenival, *Living Architecture*, 86.

21. Every Egyptian temple is unique, but the basic elements of its architecture remained stable throughout its history, reflecting geographical and cultural stability. Architectural flourishes and finely carved images never before seen, in place of the traditional, give Karnak an individual ethos, but its layout is "classic" (De Cenival, *Living Architecture*, 173).

22. "House of the morning," Spencer, "Priests and Temples: Pharaonic,"261; *Arts and Humanities Through the Ages*, 1: 48; Lloyd, *A Companion to Ancient Egypt*, II, 787, 792; "the horizon," De Cenival, *Living Architecture*, 89.

23. "Sunk relief," stone cut away from the surface while maintaining the integrity of its plane, so that a sheet of paper laid over the sunk relief would lie absolutely flat. Quintessentially Egyptian, the technique reached its maturity in the New Kingdom a thousand years after it first appeared. "Light-catcher," Giedion, *The Eternal Present*, 153–55.

24. This is true of Egyptian painting as well. In the Eye of Horus holding a torch, for example, there's no angle of light from the torch. Conversely, if no torch were depicted it wouldn't change the scene of light (see Michalowski, *Art of Ancient Egypt*, pl. 573, 413). It is the same with the solar disk as depicted with two lions bearing the disk (pl. 576). As it is also in the Nightly Journey of the Sun with light background and no effect of light in darkness (pl 387, 390). The flip side of this is in outdoor scenes, such as Nebamun hunting birds, where the light is homogeneous with no hint of directionality (see *Nebamum Hunting Birds*, Robins, AAE, pl 11, 22).

25. The experience of light-shadow within the temple, though a reminder of this cosmic boat journey, was one of the physical presence of a new kind of light dramatized in its dance: a dance engaged in daily by each individual Egyptian who could never walk in the stone-light outside the temple without his shadow. This was inseparable from his understanding of himself as an individual because his body, his shadow, and his name were all of equal weight. Though a component of the individual, it also possessed a separate existence. Hence the image of a god on a temple wall could be referred to as "god's shadow," as paradoxically could the temple itself (Allen, "Shadow"). "Light of life and the shadow of mystery," De Cenival, *Living Architecture*, 91. This catching and casting rhythm is picked up by a frieze of clerestory windows high above the priest running the length of the hall on each side of the central axis. These fill the gap between two roof levels created by the different heights of the two rows of columns to his right and left. Flanking the priest as he strolls up the central passageway of the Great Hypostyle Hall, the inner row soars seventy-eight feet. And just beyond, the outer one rises but ends below the first. The high windows could have admitted a flood of light, as they are allowed to do in later Christian basilicas. Instead, out of the need to express a specific quality of the character of Egyptian light, the architect fills them with perforated stone grilles. These mediate the light in order to sustain and heighten its light-shadow dance, interrupting shafts of light and directing diffused beams of sunlight to create constant movement below (Giedion, *The Eternal Present*, 391; Wilson, *The Culture of Ancient Egypt*, 791).

26. "He who rejoices . . . Disc,'" Redford, *Akhenaten*, 62.

27. Redford, *Akhenaten*, 57–58, 138–42.

28. "He who Sees the Great One," Hari, 9–10; Akhenaten adapted the Heliopolis sun temple's open-air courtyard (Lehner, *The Complete Pyramids*, 150). A *benben-stone* also in the courtyard, possessing great creative power derived from the sun (Wilson, *The Culture of Ancient Egypt*, 784), symbolically reinforced this relationship between priest-king and the sun (Quirke, *The Cult of Ra*, 99).

29. "Under the sun," Hornung, *Akhenaten*, 70, 73.

30. Redford, *Akhenaten*, 165; Hornung, *Akhenaten*, 26–27; "fairly ecstatic," Breasted, quoted by Hornung, *Akhenaten*, 13; "role of the sun as light," Hollis, review of *Akhenaten and the Religion of Light*, 500; Szpakowska, "Religion in Society: Pharaonic," 508.

31. "Pure presence," Quirke, *The Cult of Ra*, 157. To reinforce this understanding, Akhenaten resurrects the Old Kingdom word "Aten" signifying simply "disk," which could refer to any disk-shaped object, a mirror as well as the appearance of the sun in the sky (Redford, *Akhenaten*, 170). This move revived a dead metaphor in effect, as the word was an ancient hieroglyph of the sun's rays meaning "to shine." In its radiating the sun "calls the world to life and keeps it alive." That is, Akhenaten declared himself independent of the Amun-Re temples at Karnak in the south and of the authority of the sun temple at Heliopolis in the north (Anthes, "Mythology in Ancient Egypt," 46; Hornung, *Akhenaten*, 18, 53–55, 62, 93).

32. Redford, *Akhenaten*, 169–70, 172, 174–75, 180.

33. "Mythic dimension" and "aloneness," Assmann, *Egyptian Solar Religion*, 67–68; "light radiating from the sun," Schlögl, *The Ancient Gods Speak*, 23. Thus Akhenaten created perhaps the world's simplest "religion," no secret temple, no revelation from on high, no scriptures. Yet one conceived as universal, valid beyond the Egyptian people and their land, its devotion to Aten being both more and less than a religion.

34. Whatever Akhenaten's motives for establishing a new starting point in the Egyptian experience of light, we are confronted with this new relationship. Whether longing for the beauty of light, whose icon is the sun-disk, or needing to reinforce his status and power as priest-king by identifying with the sun-god, possibly out of insecurity in his court, the individual bathed in sunlight remains (Redford, *Akhenaten*, 178–79).

35. "You appear beautiful . . . / sparkling," "transfiguredness," Assmann, *Egyptian Solar Religion*, 74, 94; "fine gold," The Egyptian Book of the Dead, pl 20, chapter 15; c. 1250.

36. "Natural," Assmann, *Egyptian Solar Religion*, 73; "the lord of sunbeams, maker of brightness," Redford, *Akhenaten*, 177.

37. "The Living one, . . . the light that comes in the sundisk," quoted by Allen, "The Natural Philosophy of Akhenaten," 90–92.

38. Plas, "Hymns"; Assmann, *Egyptian Solar Religion*, 28, 36.

39. "At dawn without failing . . . eyes of the herds," "rises at dawn without failing" (Lichtheim, Ancient Egyptian Literature, II, 87). A Vedic hymn to the dawn from about the time of the New Kingdom in Egypt parallels Suti's: "She dispels the great gloom to bring the earth to view, / . . . she makes the light. / The sun, . . . ascending / joins her in scattering the light. . . . Thou makest the earth visible" (Bose, *Hymns from the Vedas*, 93, 95).

40. Redford, *Egypt, Canaan, and Israel in Ancient Times*, 388. The fundamental difference here between the Egyptian and Hebrew notion is that for the latter, god and light are separate though occasionally, because of foreign influence, represented as a winged sundisk, emphasizing his role as all-seeing judge (Wolde, " In Words and Pictures," 265–67). Nevertheless, for the Hebrews light is distinctly a creation of God, not self-created. This doesn't preclude an aesthetic response, however, such as Qohelet's, "How sweet is the light, what a delight for the eyes to behold the sun!" (Eccl 11:7; trans JPS), but he is in the minority, being given to sensual experience. More typical is God's rhetorical, almost mocking question to Job, "Where is the way where light dwelleth" (38:19). Job knows the answer; without God it doesn't exist. In Hebrew experience light is strictly subordinated to God. It is "not somehow an overflow of the essence of deity, but rather an object, even though preferential, of God's creation" independent of the celestial bodies, which in Genesis (c. twelfth-century BCE) were created two days after light (Rad, *Genesis*, 49). Accordingly, light acts independently of the sun, moon, and stars to dispel darkness. Only in this sense, as in Akhenaten's apprehension of light, it is anti-mythical.

41. "Built himself with his own hands" quoted, Hornung, *Akhenaten*, 84; "shining in your form of living Aten," quoted, Assmann, "Akhanyati's Theology of Light and Time," 147–49.

42. Quoted, Assmann, "Akhanyati's Theology of Light and Time," 150–51. Henceforth, all quotes from Akhenaten's "Great Hymn" are from Lichtheim (Ancient Egyptian Literature, II, 96–99), unless otherwise noted.

43. "Central principle," Hornung, *Akhenaten*, 93; Smith, *The Art and Architecture of Ancient Egypt*, 177. Akhenaten's emotional experience of a naturalistic light opens up the visible world, revealing its human particulars. A greater awareness of man's humanity, his life in the world and in time, enters the hymn (Lichtheim, Ancient Egyptian Literature, II, 98: Quirke, *The Cult of Ra*, 160). Due to this opening up, in art Akhenaten finds new expression and introduces a new style, which for the first time in Egyptian art depicts interactions between figures, such as

between him and his wife, Nefertiti, and their daughters. Hand-picked artists such as Bak, Akhenaten's royal sculptor, who were instructed by him in the new style, express a range of emotions in these interactions, from the joy of mutual affection and ecstasy of receiving the touch of the sun's "hands" to grief at the death of his second daughter, Meketaten. His artists achieve this expressive power by creating a different space for their figures, which has a more open three-dimensional effect than before, and depicting naturalistic details never before seen, such as the king's individual toes (Freed, Pharaohs of the Sun, 116, 118, 121). His contemplative apprehension, at once rational in its insistence on the visible and personal in its intimacy, inevitably precipitated his breaking with tradition.

44. "Makes seed grow . . . still his tears," "When the chick . . . comes from it," "far sky . . . river's course," Lichtheim, Ancient Egyptian Literature, II, 97–99.

45. "Communicative," Assmann, "Akhanyati's Theology of Light and Time," 152–57.

46. "To become, to come into being, to develop," Assmann, "Akhanyati's Theology of Light and Time," 155.

47. "You are in my heart, . . . your might," Lichtheim, Ancient Egyptian Literature, II, 99.

48. "Divine energy," Assmann, "Akhanyati's Theology of Light and Time," 159–60; "There is no natural principle but light, and no god but Akhenaten," "great Queen . . . living forever," Allen, "The Natural Philosophy of Akhenaten," 100.

49. "God is nothing else but the sun . . . nature," Assmann, "Akhanyati's Theology of Light and Time," 161. It may seem for a moment that Akhenaten has re-mythologized light. But this is not the case, as is clear from the testimony of his "Great Hymn," the dramatic account of his personal encounter with light from which a new "revelation" emerged. Visible light, energetic author of time and space, is a function of the sun's radiance. Nothing exists apart from light; it is the "One," which for Akhenaten is more a physical than theological notion.

50. Assmann, *Egyptian Solar Religion*, 144–45, 174, 188.

51. Hornung, *Akhenaten*, 96.

52. The Egyptians adopted this mythology from neighboring cultures such as the Babylonian, where light and darkness do battle and an adversarial relationship dominates. In the Babylonian epic of creation, the *Enuma Elish* (c. twelfth century BCE), Tiamat, primeval darkness, struggles with Marduk, "son of the sun," Gray, *The Mythology of All Races*, VI, 263. In the Babylonian *Epic of Gilgamesh* (c. 1800–1600 BCE), contemporary with the end of the Middle Kingdom in Egypt, Gilgamesh manages to survive this darkness as he traveled the "Road of the Sun" on its journey through the netherworld of night (Gray, IX, 227) one league, "dense was the darkness, light there was none" (l. 229), two leagues, then twelve with the same refrain sounding. But at last he "came out before the sun(rise)" and "it grew brilliant" (ll. 278–79); and see Kovacs, *The Epic of Gilgamesh*, 77–78.

53. One must shun evil on earth to avoid the underworld of darkness and merge with Re in heaven, realm of light. To ensure this future the Egyptian priests provided a guidebook, the *Book of the Dead*, which was buried with the mummy of the deceased. The most popular chapters, inscribed on coffins and the walls of tombs and temples, displayed a profound uneasiness about the darkness of the netherworld and a desire for safe passage (Lesko, "Book of Going Forth By Day"). The book opens with a poignant expression of the fervent wish to be accepted in the presence of Osiris, Lord of the netherworld of darkness, and to join Re in his boat and live forever with him in the celestial world of light. A royal scribe of Ani, overseer of the granary of the Lords of Abydos, makes his plea, "I have taken hold of the bow-warp of the Night-bark and the stern-warp of the Day-bark. May he [Osiris-Re] grant that I see the sun-disk; . . . may a place be made for me in the solar bark on the day when the god ferries across" (Faulkner, *The Egyptian Book of the Dead*, pl 1). Osiris-Re is pictured on this page of the *Book of the Dead* as rising above the horizon from between the boats with the sun-disk in his hands, while six baboons lift their hands in praise, being the Egyptian equivalent of roosters, as they do in everyday experience (Goelet, Introduction, 154). Generally available to anyone who could afford it, by Peta's time even women owned this book and may have composed parts of it.The Vedas of the Indus Valley, which were composed roughly at the time of the New Kingdom in Egypt, express a dualism born of anxiety: "Will the Sun rise again? / Will our old friend the Dawn come back again? / will the power of Darkness be conquered by the God of Light?" (Mason, *Light Fantastic*, 9). They also identify light with immortal life and goodness;

the Vedic poet says, "We have drunk Soma [an hallucinogen]; we have become immortal. We have gone to the light" (Campbell, *Masks of God: Oriental Mythology*, 181; for the progression from an adversarial relationship to dualism, with its ethical valence of justice and injustice, good and evil, see "Light and Darkness," *Encyclopedia of Religion*, 8: 5450–51).

54. Dresden, "Mythology of Ancient Iran," 341.

55. For this reason the period has been commonly labeled the "Axial Age." In this age the identification of the divine and human in the king is broken (Robert N. Bellah, *Religion in Human Evolution: From the Paleolithic to the Axial Age* [Cambridge: Harvard University Press, 2011], 232), and the main question becomes: Who is the true king, the one who "reflects divine justice" (266). It was a time when challenges to the dominant cultural order became "widely apparent." These involved social criticisms that came from "renouncers," such as the Buddha, the Hebrew prophets like Amos, and the Greek philosophers like Socrates (573–75). See Robert N. Bellah, *Religion in Human Evolution: From the Paleolithic to the Axial Age* (Cambridge: Harvard University Press, 2011), chapters 6 to 9; and Karen Armstrong, *The Great Transformation* (New York: Anchor, 2006). However, the definition and unity of the period are problematic. See Jan Assmann, "Cultural Memory and the Myth of the Axial Age," in *The Axial Age and Its Consequences*, edited by Robert N. Bellah and Hans Joas (Cambridge, MA: Harvard University Press, 2012).

56. "Light radiating from the sun," Schlögl, *The Ancient Gods Speak*, 23.

57. His is a different "religion," indeed, because informed by an existential encounter with light. This is evident from his overt acknowledgement of complete identification with the Aten, which we remember is "light radiating from the sun." This is signaled early in his reign when he takes a new name. Changing from his given Amenhotpe to Akhenaten, he adopts a name that means "illuminate manifestation of Aten" (Schlögl, *The Ancient Gods Speak*, 23–24). Forking paths, Allen, "The Natural Philosophy of Akhenaten," 100.

Chapter Four

Scenes of Light and Shadow, the Luminaries

The Greek Classical Age

Standing before a pyramid, an Egyptian priest saw reflected from its polished limestone skin the gleam of divine light. How different, centuries later, was the experience of Thales (620–546 BCE) as he stood before the same pyramid transfixed by its shadow etched on the desert by a brilliant Egyptian sun. This experience proved to be central to Greek culture and pivotal in the human history of light. In fact, Plato named Thales the first Greek "sage" or wise man, what I call a "Luminary," revered for his desire to know, his curiosity, not for his function as priest or king. And these men, primarily of Greece (also of India), shared life-changing encounters with light. Akhenaten knew this experience, of course, but his residual religious awe limited what light could reveal of its character.

The Luminaries, men like the philosophers Thales, Parmenides, and Plato, and the playwright Aeschylus, as well as the scene- and vase-painters, and architects of this chapter, all participate in the colloquy of light and shadow, revealing hitherto unknown aspects of light's character. Neither complementary nor adversarial, as we'll see, light and darkness enter a new relationship. Thales addressed shadow first and light joined, whereas Parmenides turned to light first and shadow joined. And no sooner does Aeschylus bring light on stage than the character argues with shadow, whereas Plato in his dramatic dialogues brings shadow into the conversation as a measure of the soul's progress towards the "pure light" itself, which he exalts as the very "substance" of the universe.

THALES' LIGHT AND SHADOW ADVENTURE

Thales' profound experience of light in its sharp contrast with shadow forms the axis of the Greek experience of light. And he is justly considered the founder of a school of the natural philosophy glimpsed by Akhenaten, whose hymn he may have heard on one of his trips to Egypt from his home in Miletus. The boy-king broke with tradition in considering light as light, the self-contained creator of time and space, and sustainer of life. Yet his "Great Hymn" is also a narrative reinforcing his identity with a god. His Greek heirs like Thales, however, as we move from the Egyptian priests to the Greek sages in this chapter, stand wholly apart from the gods. Not only do they consider nature, including light, a phenomenon in itself, worthy of attention, but of study as well. It is in this tradition that light became for the first time a "subject-matter."

So in the sixth century BCE, some eight hundred years after Akhenaten's death, Thales' radical step, finalizing the divorce of the sun from the gods, allowed him to experience it in new ways, placing a value on systematic and precise observation. Entranced by a pyramid, Thales becomes obsessed with the relation of light and shadow. He observes that the pyramid functions as the shaft (called a "gnomen"), in effect, of a large sundial. And standing before it, he discovers ratios, his shadow being directly proportional to the pyramid's. The triangles formed by its height and length of shadow cast on the ground and his body's height and shadow are related in exact, not mysterious ways. Thales, that is, reads what the sun "writes on the ground."[1]

He realizes that the light from above inscribes on earth below information, these ratios, about the structure of the universe, expanding his experience of light and shadow. Thus the gnomen "observes" in a sense and reports the results by tracing the boundaries of its shadow on the ground. It is a tool of discovery. And its report on the unseen, which is knowledge in and of itself, is inseparable from Thales' direct experience. On the one hand, this involves the senses, namely the eye; on the other, the pyramid as gnomen dispenses with the senses. For out of the light-and-shadow scene emerges a geometric model that not only has earthly use in enabling him to measure the height of the pyramid but is the basis for a theory of the cosmos. The pyramid as an instrument, a gnomen, casts a shadow that reveals more than itself.

Thales' experience represents a new consciousness in the world that yields a new experience of light, which in turn informs the new consciousness. We witness a seismic historical shift that introduced "thinking about thinking," an analytic set of mind layered over the ancient mythic consciousness that had dominated the previous ages of shaman and seer as well as priest. We saw glimmers of this in Akhenaten's hymn, but Thales took a radical turn in our cultural evolution, which the tragedies of Aeschylus that we'll visit later in this chapter consolidated in their bringing a Dionysian light on stage. Likewise, the drama of Plato's dialogues rises from and furthers this consciousness, as does the diagrammatic expression of light in Euclid's treatises.[2]

Cultivated and developed by these Luminaries, we'll see this new "theoretic consciousness" extended in the next chapter as well where we meet what I call the "Refractors," such as Lucretius, Plotinus, and Augustine, interpreters of the Luminaries. Experiencing light in new ways, then, the Greeks discover new aspects of its character, such as its contrary relation to shadow, and the Romans its history, a "biography." Together these chapters explore the Classical Age of Athens and Rome, respectively, with a glance back at Homer.

ATHENIAN LIGHT: THE POTTERS' LIGHT-SHADOW CONTRARIES AND HOMER'S RADIANCE

Changing the human relation to light, the new consciousness was fostered by a critical invention, the Greek city-state or *polis.* A walled sanctuary with an open space (*agora*) where citizens in the light of day, which imposed obligations of truth, gathered to talk and trade, the *polis* encouraged discussion of ideas. And this mode of interaction was consolidated by a powerful technology of memory, the alphabetic method of writing, far more flexible and easier to master than hieroglyphics. Enabled by these inventions, the sage gains a vantage point, a place to stand outside his society and his received notion of the cosmos and the divine. Such a stance allows him to reflect on his experience of light analytically for the first time.

This was the light that shone on the agora of Athens. About three miles from the sea, this city sits high above it on the central plain of Attica. Because the sun shines most of the time during hot, dry summers and mild winters in a typically cloudless sky, the atmosphere has a special quality, one that Euripides notes in his play *Medea*, where he alludes to its "transparent clearness." As a consequence of this quality of light, Greek culture enjoyed a special relationship with light, C. M. Bowra writes, which had a "powerful influence" on its vision of the world. Athenian citizens spent most of their time outside in brightness. They worked, ate, talked, and played in the open air under the sun. They gathered for political and judicial deliberations in the Acropolis in the center of the city as well as for theater performances in the open amphitheater. Who they were was inseparable from the light in which they lived their lives. To appear in the light of day in public was in itself a mark of truth for the Greeks. It is in the light that Greeks know they have life itself; one becomes understandable by being examined in the light. And this extends to the physical world. Daylight, in effect, makes "existence understandable to itself."[3]

At the same time, we cannot understand the Greek experience of light, as with Thales', apart from their experience of shadow. For the sages these phenomena exist in a contrary relationship; shadow imposes a limit on light that in turn tests the limit. Neither negates the other, but each defines the other. Neither being independent of the other, they exist in a relationship of mutual questioning as if in the *agora* debating. Making sight and the limits of sight (blindness) possible, this becomes the ground of their existence in which mind and world have their unity.[4] We've seen the Egyptian light-and-shadow dance where light and darkness are complementary, each taking the lead alternately, and later, antagonistic, each battling from a moral stance of good or evil. And these relationships were revaluations of the interpenetrating light and darkness in the experience of the prehistoric shamans and their seer successors. For the Greeks, however, this relationship is neither complementary nor antagonistic but contrary.

This contrary relation defines Greek experience. For light outlined, which is shadow outlined, think of the shadows cast by tree trunks on snow creating bands of shadow alternating with bands of light, each limiting the other. An image that for the Greeks comes from nature via something like the gnomen, as we've seen, or from their architecture, such as temple pillars, this is the core image of its culture. And "image" is an "act *of* consciousness," as the French philosopher Gaston Bachelard reminds us, "not a thing *in* consciousness," hence possessing the power to make *uncommon* in our experience what Reverend Hunter calls the "common" everyday light.[5] Thales' uncommon, personal experience of light and shadow as he stands before the pyramid, which is provided by the clear atmosphere of Greece producing in the landscape firm outlines, exposes the heart of Greek culture and its new theoretic consciousness. Probing its cultural imagination, then, we find the contrary relation of light and shadow, and have arrived at the bedrock of the Greek experience of light, as basic as the great limestone massif on which Athens rests.

It is not the politicians or even the first "philosophers" of the period, a term the Greeks didn't use until some two hundred years later, that give the fullest expression of this image of shadow defined by light defined by shadow, but the potters and painters. These artists arose in the sixth century at a time when the Greeks were consolidating a sense of themselves as Greeks and individuals, made possible by the rising wave of theoretic consciousness, which in turn was intensified by them. This is no more evident than in the artifacts of individual potters and painters of Thales' age. We know almost nothing about their social status or training, but we do know they were intensely competitive, as were contemporary dramatists, such as Aeschylus, and so a number signed their works. Conscious of experimenting, these potters and painters, the artist being both sometimes, were very aware of themselves as innovators and individuals.

Their painted vases, while employing inherited artistic conventions, then, often express a personal experience of light and darkness. This was a relationship of contraries that found its first full expression in the individual artist's outline drawing on pottery vases sold widely to the Greek citizenry, from large amphorae to modest drinking bowls. Known as black-figure and red-figure painting, this style commonly featured mythological or Olympian heroes painted black on a light ground or vice versa. In either, figure and ground enter into a dialogue. And the artists exploited the dramatic contrasts of this figure-ground reversal, which shaped the experience of countless Greek citizens over some three hundred years.[6]

Light on dark or dark on light, red-figure painting being simply a reversal of earlier black-figure, both modes exploit the contrast between glazed areas, which are fired black, and reserved unglazed areas, which take on the lighter hue of the clay on firing. What remains constant is the "silhouette principle." This is because, as Thales would have seen immediately, black-figure painting was derived from earlier shadow-painting (*skiagraphia*) in which the shadow outline cast by a figure, such as a horse, is recorded. Such a principle finds its supreme expression in black-figure vase painting. In this technique, which began in the seventh century, the artist engraved his lines with a sharp tool, and the figure or decoration appears as a black silhouette on a reddish background. In red-figure painting, introduced by the end of the sixth century and lasting into the later fourth, he outlined his figure in black and set its silhouette with a black background left unpainted, the figure showing through in the reddish tones of the fired clay. The emotional force of either painting lies in his exploitation of the silhouette principle.[7]

Light is defined by shadow, as shadow is defined by light. The black-figure painting insists on the medium of light in which its figures often float, suspended (Figure 4.1). The later red-figure insists on the void, a hard and opaque jet black glaze surrounding the figures, a ground uncompromising in its starkness (Figure 4.2).[8] Yet the reddish figures themselves in contrary relation to the black glaze are vibrantly luminous. Both techniques rely on figure and ground, light and shadow interdependence. The artists dramatize this by deploying open and closed negative spaces that argue visually among themselves. In both black- and red-figure painting we see the artists' emotional encounter with light and its contrary shadow as they manipulate their ratios.

This refinement of shadow-painting thus became in the hands of the vase-painters a language to express light's contrary relation to darkness and ultimately the Greek consciousness. They shared with Thales the new apprehension of light's character. And though an individual expression, this new expression resonated with the experience of citizens throughout Greece. Black- and red-figure painted vases, native to Greece alone, were immensely popular, and anyone with a modest income in the Greek city-states could

Figure 4.1. Sappho Painter (5th century BCE). Attributed to. Terracotta lekythos (oil Flask), c. 500 B.C. Archaic Greek, Attic. Terracota; black-figure, white ground. Image copyright © The Metropolitan Museum of Art. Image source: Art Resource, NY

Figure 4.2. The poets Alkaios and Sappho, Greek red-figure wine cooler, c. 475 BCE *Bibi Saint-Pol (photographer). "I grant anyone the right to use this work for any purpose, without any conditions, unless such conditions are required by law." Accessed 3/15/17. https://commons.wikimedia.org/wiki/ File:Alkaios_Sappho_Staatliche_Antikensammlungen_2416_full.jpg*

afford this pottery for their tables.[9] Individual styles were recognized and coveted, each artist varying the way he massed blacks or reds against their background. But the fundamentally new experience of light embodied in these vases became the Greek experience. Despite variation and innovation in early black-figure and its mirroring later in red-figure painting, this basic expression of light and darkness as contraries seized the imagination of the citizenry in all areas of the Greek world. In fact, it became so much a part of Greek consciousness that it was identified with consciousness itself in the Greek theater, as we'll see later.

But what is the contrary relation of light and darkness these vases express? They document a light independent of shadow, shadow independent of light, and not. Each limits the other, dark and light areas being skillfully interrelated so we perceive them as fundamentally interdependent. This experience of light and darkness perhaps reflects the tension between a man's liberty as an independent Greek citizen and his constraint as a member of the polis joining his fellow citizens daily in trade and debate. Each calls the other into question.[10] The Greek as individual and citizen, then, moves daily from darkness into light and from light into darkness literally and metaphorically. And this in turn calls to mind life on earth in the light passing inevitably at the end into darkness. Thus the objective of the highly stylized vase-painting, black or red is quite possibly not to depict a human or mythological figure realistically but to pass into and out of death, of which his daily exiting and entering shadow may on occasion be a reminder. For every Greek knew his soul was a "shade" (*skia*) or shadow on earth in conversation briefly with light, which after death would return to the underworld of shadow.

Always in dialog, as are death and life, neither light nor shadow can be separated from the other. This relation is dramatized on a black-figure oil flask (lekythos) used for funerary offerings that depicts Helios (sun), Nyx (night), and Eos (dawn), as well as Herakles (c. 500–490 BCE) (see Figure 4.1). The artist's personal vision bursts through the conventions of vase-painting here as he presents us with a scene that is uniquely cosmic yet detailed.[11] The central image is the sun's orb, a round disk with rays, not the mythological figure Helios standing his chariot horses just below. The artist calls our attention to the sun itself, even showing us Helios' hair catching its light. But what arrests us is the orb's force field in which night (Nyx on left) and day (Eos on right), darkness and light, appear to be suspended. Flanking the sun, they are drawn in opposite directions yet seem to be held by the gravitational attraction of the sun, ground of their existence. Caught in a field of attraction/repulsion between the opposing curves of night and day, as Nyx and Eos drive, respectively, to the left and right, we feel the tension of light and shadow in the artist's experience. Dependent on each other, neither of these elemental forces dominates the other.

Darkness and light mutually limit each other and by doing so, I've suggested, define each other. When Thales' contemporary, the poet Alkaios (see Figure 4.2), entreats the gods for protection, "light a clear path through the midnight gloom / for our black vessel," he speaks for all Greeks of life itself. And when the poet Sappho, also a contemporary (pictured on the right), speaks of the moon's light that "stretches out over the salt- / filled sea," she is poignantly aware that the moon hangs in a black void. So the figure of Sappho stands in a void on the vase. And Alkaios' lyre holds a chunk of blackness in its body even as its song lights up Sappho. Light outlines and limits this heart-shaped blackness even as blackness limits while contouring the beautiful body of the lyre in a visual dialectic. Infused with the contrary drama of light and darkness, this scene dramatizes not only the sweet interaction of lovers but that of light and outline, life and shadow.[12]

That this experience of light-shadow contraries arose with the theoretic consciousness is evident when we go back in time some three hundred years to Homer. The contrast is stark. Not light and its limits but its radiance alone impressed itself deeply on the hearts of the ancient poets. The gods live in the "white radiance of sunlight." They delight for eternity in the bright, cloudless atmosphere of Mount Olympus like the light, not so strangely, that bathes Athens. This harks back to the Indo-European word for "light" that refers to what is "brilliant" based in the idea of free, clear space in the open air. And the word Homer uses for "white" (*leukos*) refers not so much to whiteness itself as to its intensity, its brilliance. Not only do the gods dwell in brilliance, they are radiant in themselves. Radiance for Homer is divinity. When a god appears in an epiphany to a Homeric hero, their calling card is radiance.[13] "Flashing earthward" (*Iliad*, 4: 78), Athene appears to Achilles, and he recognizes the god by her "terrible eyes shining" (1: 200). Ecstasy and fear come with this divine epiphany; such is the power of radiance.

Homer's experience of radiance as a mix of these emotions was influenced by earlier expressions in Mesopotamia and Egypt. His epics exhibit a mythic consciousness that assumes a model of the universe inherited from earlier tales. As a consequence, once he has recovered from the initial experience of the Mediterranean sun's radiance, an epiphany, Homer identifies it with divinity. Occurring throughout his poetry, it appears at critical moments, such as with Athene's appearance, and even orders the larger movements of his poems as he relates the experience of heroes such as Achilles. In doing so, Homer constructs a cosmology from inherited fragments of older myths that requires radiance. For it becomes in his experience the primary image of earthly transcendence. He projects radiance into the realm of the gods where darkness erases the light of life, having been on earth simply Night the complement of Day, never its contrary.[14]

It is interesting that while Homer was singing his epic verses in Greece, nomads in India were reciting the sacred words of the 1,028 hymns of the

Rigveda made by poets some five hundred years earlier who called themselves "Children of Light." As with Homer, their deepest experience of sunlight is radiance, the sky itself being "flowing streams of light," which were said to be the poets' vision of "ultimate reality." But where Homer felt himself embedded in a radiance external to him, the Vedic poet who shared his mythic consciousness felt this "dazzling light" wound him internally. And this existential experience of radiance precipitates an inward change in his way of being. So the Greek poet looks outward to radiance, whereas the Indian poet looks inward to the pure light, the Absolute (Atman-Brahman). Radiance lies at the heart, then, of both the earliest Indian and Greek experience of light's character.[15]

Homer blurs the line between the radiance of natural light and divinity, the literal and the figurative. Is the image of light that he perceives, such as a blaze of sun at noon, literal or symbolic? The ancient poets in Greece and India did not distinguish them within their mythic context. Furthermore, they did this unconsciously. We see a few centuries later with the advent of the theoretic consciousness and the sages of the Classical Age, such as Thales, however, that they did this consciously, fully aware of the distinction. This movement from image to metaphor, and to symbol, will accompany us throughout the rest of this book. Those having specific encounters with the character of light, the initial natural *image* as directly perceived and imagined, make these existentially meaningful through interpretation, *metaphor*. Yet the natural phenomenon is virtually inseparable from metaphor, being at once purely literal, pointing to the original image of light, and aesthetic, being a personal interpretation. And many consciously extend this interpretation, often transforming their experience of light into *symbol*.[16]

Homer experiences the "white radiance of sunlight," what I call the "core image" of his encounter with light, in an ordinary literal sense. But he bends this perception like a blues note into a figurative meaning, as when the light of dawn (literal) becomes "rosy-fingered dawn" (metaphor). Homer invests his experiential image of *radiance*, then, with valuations rooted in personal desires. So his experience of radiance, a "literal" image that is the direct outgrowth of the body and psyche just as the eye is of his brain, takes on a specific emotional charge, which he expresses in "metaphor," a move familiar to us in Akhenaten's hymn. Furthermore, "radiance" comes to be "divinity," Homer's image becoming ultimately a symbol of the sacred. Light is divinity manifest, that is, literally *and* figuratively as it is given a personal meaning, which derives partly from Homer's cultural inheritance, partly from his individual fears and desires.[17]

What sets the later sages apart from Homer, however, is their consciously distinguishing, as inheritors of the new theoretic mindset, between the literal and symbolic. They move self-consciously by extended interpretation from the literal image, bound up with metaphor, beyond the aesthetic to the symbolic, which came to dominate the theoretic consciousness of the Classical Age. The sages leverage the power of this symbolic consciousness, giving fuller and more precise expression to what Reverend Hunter experienced as the "glorious" (aesthetic) and "mysterious" (symbolic) aspects of light's character. This conscious experience both ushers in and bolsters theoretic consciousness, as our relatively unconscious dealing with a "common" and "simple" light in earlier ages did not. Plato, as we'll see later, wrestles with the sun as image, whose meaning is existence, and as symbol in which meaning detaches itself from mere existence. And the light and limit given by the sun, Thales knows, is not wholly literal or metaphoric, but rather a symbolic tool that he can use to probe nature.[18]

If the personal language of metaphor is strong enough to be assimilated by conventional language, such as light being the world of mind, it becomes the expression of a community's experience. Metaphor remains grounded in physical experience because the image, such as high noon's sun-dazzle, that crystallizes it is infinitely complex and open-ended, thus resisting reduction to any one metaphor.[19] But though they begin and end in sensory experience, metaphors eventually die as clichés, as did Homer's "rosy-fingered" dawn.

Enter a new image. Bursting from a new experience and the pressure to express this, the image gives birth to new metaphor. The movement from the perceptual (image) to expressional (metaphor) to the conceptual (symbol) is thus circular. One feeds the other within a culture as image feeds metaphor, which feeds symbol, which, in turn, shapes experience. We all interpret our individual experience, of course, in terms of our cultural inheritance, which in turn is shaped by our personal experience. Such reciprocity of the cultural and the personal allows an opening for the seed of a new consciousness and with it a new valuation of light to be planted. We'll see this gyring through time as we move from Parmenides, a younger contemporary of Thales, to Plato in this chapter and to Neoplatonism in the next, following the successive unveiling of light's character.

PARMENIDES' REFUSAL TO NAME LIGHT AND NIGHT

Parmenides develops the symbolic transformation of his core experience of light in such a compelling way that Greek culture absorbed it to its very marrow. A new kind of man, the itinerant philosopher from the colony of Elea was the kind that Sophocles described in his *Antigone* in 442 BCE, a confident human in possession of a "wind-swift motion of brain." In 450 at about age sixty-five, his "hair almost white," he visited Athens. And when Plato met Parmenides, he found the Eleatic's arguments so compelling that he adopted his idea of an unchanging reality behind all the flux of the world.

More important, he adopted his method of extended logical argument. Parmenides was the first to employ a rigorous method hinted by Thales in which the gods had no place, giving the theoretic consciousness in his dogged thinking about thinking a firm foothold.[20]

Just as Thales had been entranced by the sun's light and shadow, Parmenides was captured by the moon's light and shadow. But where Thales' relationship with light and shadow led him outward to the cosmos, Parmenides' relationship, a more complex one, directed him inward to the mind itself. From careful observations he learned that the moon was a sphere and realized that it could be used as a tool of intellectual discovery, just as Thales before him had come to see that the pyramid was, in effect, a giant gnomen and could be used in a similar way. Parmenides' core experience of light was that of the moon's light-dark cycles, then, which he carefully observed night after night, concluding that the moon receives its light from the sun and arriving at the true explanation of the moon's phases.

But he was shaken by this discovery. How could it be that the moon's alternating light and dark, its periodic waxing and waning, is merely light and shadow play. Unsettled, he had been persuaded that the very substance of the moon changed. So he felt duped by the shifting light and shadow, which had initially convinced him change is real, a notion his faith in the order and constancy of the cosmos could not accept. His great discovery of the cause of the phases of the moon, that its waxing and waning was only appearance, in fact, "shocked and overwhelmed" him at first.[21] What his senses told him contradicted what he "knew" to be the truth. Unlike Thales' confident relationship with light and shadow, Parmenides' is fraught, blinking with ambivalence. On the one hand, light and shadow lead to truth; on the other, they conspire to deceive. But how to escape the state he finds himself in?

His way to emotional and intellectual relief was to write a philosophical poem in which he traces the path to a resolution of his dilemma. "The Way of Truth," part one of his poem following a visionary prologue, is an "argument" that leads to proof that despite the moon's changing appearances, change is not real. "The Way of Opinion," part two, rehearses the appearances of light and shadow that had duped him into believing that change is real. Putting words in the mouth of a goddess who symbolizes philosophical inquiry and takes him in her chariot on an intellectual journey, Parmenides proves that pure sensory observation is self-contradictory because it leads to illusion (part one), namely, his conviction of the reality of the play of light and shadow. Then he returns to the observations of the moon's appearance that had precipitated his crisis (part two), but which no longer present him with a contradiction because he now knows the truth of things. Namely, it is only the moon's body itself that is reality. The apparent contradictory ways of light and shadow, then, resolve as we'll see, into contraries that admit an unchanging moon.[22]

Parmenides presents his journey as an allegory in which he travels from the everyday world to a place outside it in the intellect. Having been taken up by the "daughters of the sun" and having passed through the "gates of the ways of Night and Day," it is here that the goddess presents him with a logical method that enables him to arrive at the truth of night/day contraries. Yet his allegory feels very personal, as we'd expect triggered as it is by a crisis. He found himself as the result of careful observation of light and shadow, after all, desperately attempting to reconcile his conviction of permanence with his perception of impermanence. He is torn between his true discoveries arrived at through patient and careful nightly observations of the moon's alternating light and shadow, and his true intuition that the cosmos does not change. Parmenides is caught in a contradiction between his empirical and intellectual perceptions. And this is why the goddess gives him method, the way of contrary truth. It is a way to go beyond the senses, enabling a clarifying description of what had initially led him into false belief about things. In short, it is a way to salvage the senses.[23]

His metaphorical journey to the goddess is a mental as well as physical journey taken to resolve a personal dilemma. He is determined to discover the true relations of light and shadow and of theirs to the body of the moon. Intellectually, he takes a deductive route of reasoning into the "unshaken heart" of truth; physically, he traces a habitual route to the top of the Eleatic headland secluded "high in the air" from where he very likely made his many observations of the moon's phases. The place was sacred ground, far from the "beaten track of men," as he says.[24]

He traces in his allegory, then, both his inward and outward journey to understand why he needs the goddess' gift of the logical deductive method presented in "The Way of Truth" (the argument of part one, a watershed in the history of philosophy). By this he is able to convince himself that he can trust his senses after all. His experience of light and shadow is an encounter, in fact, with "reality." Night becoming by phases day, reflected light, and day becoming night, cast shadow. But being itself, the all "which is" does not change. The unchanging "Thing in itself" is the moon on which the illusory dance of day and night is projected. And so his journey leads him back to where he started, on the headland observing the moon's phases, the play of light and shadow.

Parmenides can trust the senses after all, his observations of these phases, yet he remains ambivalent about their interpretation. As a result, he proceeds carefully in "naming" or interpreting the now trustworthy empirical data he receives. Observation is interpretation but these acts are yet separate. His previous skepticism of the former has become suspicion of the latter, a dimension of his experience of the moon's light (its day) and shadow (its night) on the headland that Parmenides also dramatizes in his poem. The goddess instructs him in right interpretation. For she makes clear that what is "named" by us in our gullible observation of the moon's phases, its play of "light and night," as opposites is mistaken. By impetuously naming, we harden them into "principles." But this "plausible ordering" of them as opposites is a delusion. To name them, as she shows him, is to falsely interpret them. The very act of naming, shunting light and shadow into mutually exclusive categories and thus making them into negating principles, is error.[25]

If we don't name them "light and night," hardening them into "principles," in fact putting them in this mutually negating relationship, then, they cease to be opposites. And light and shadow enter into a new relationship in which they each maintain their integrity yet remain grounded in reality, the material moon. That is, they have become contraries. Such a discovery is worth many cold nights of careful observation. The contrary "ways of Night and Day" allow him in the end to retain observation, which after long and patient nights he is emotionally attached to and feels the truth of in his heart. And he can retain method, the way of right interpretation, which after long and patient thinking he is grateful for and knows leads to truth. Observation alone misleads, he knows, but observation with method restores his faith in the senses. Thus by proper interpretation he preserves the contrary relationship of light and shadow.

Both Thales and Parmenides imagine light and darkness as contraries, though approaching them from opposite angles, giving a privileged place to light or to darkness, respectively. Thales discovered contraries; Parmenides justified and defined them precisely. The very fact that Parmenides tells us in "The Way of Opinion" that "all things have been named light and night," only to make clear this is an act of false naming, reveals that he means to overcome any dualism of light and darkness. If we don't name them, avoiding the trap of convention that Heraclitus (fl. c. 500 BCE) fell into with his dualism a couple of generations after Parmenides (flourished 540–480 BCE), then they are not "essentially" opposites.[26] The play of light and shadow on the moon is the theater that made possible Parmenides' realization that these are contraries and his discovery that the moon itself was the reality behind the show. The light-and-shadow moon is a drama of reality and appearance, a drama that occupied later playwrights and philosophers alike.

AESCHYLUS' ACTOR LIGHT AND THE LAW OF LIGHT AND SHADOW

With the defeat of the Persians in 481 BCE came a new pride in the accomplishments of man. This was reinforced by a new prosperity and the power that accompanied it, the beginning of what is commonly called the High Classical Period of Greece, roughly from Xerxes' Persian invasion of Greece to Alexander the Great's invasion of Persia (333–323 BCE). In this period confidence in the human intellect takes center stage. Shouldering aside the gods, man is, in the words of Protagoras of Aldera (490–420 BCE), "the measure of all things." And Thales' adopted city of Athens becomes the heart of this rare time in history. The country was, in effect, an Athenian empire whose flourishing trade generated the riches that made democracy as well as amphitheaters dramatizing a new relationship with light possible.[27]

Athenian theater featured tragedies by Aeschylus (c. 525–456 BCE) and Sophocles (c. 495–404 BCE) among others. Immensely popular, attended by fourteen to sixteen thousand people, the performance of these plays was a major state occasion. The celebration lasted for three days with three tragedies presented each day, such as Aeschylus' *Oresteia* trilogy, followed by a comedy (satyr play) celebrating Dionysiac abandon. The city's greatest cultural achievement, tragic drama called everything into question and fostered thinking about thinking. Hence Athens was as Pericles dubbed it, "the education of Greece."

The public theater's arena of free contemplation, a field of play in the open air under the Mediterranean sun where neither danger nor trade prevailed, not only nurtured the "discovery of the mind," but engendered a new relationship with light that became the experience of a public at large. This was possible because the tragic theater encouraged the questioning of traditional ideas, including inherited ways of relating to light. When Aeschylus and Sophocles portrayed onstage a profoundly confused character in need of understanding, people began looking at themselves. Out of this comes not only the invention of moral philosophy as we would expect, but of perspective and chiaroscuro expressing a new relation to light.[28]

Dionysiac Light Takes the Stage

In the open-air theater, light as much as the actors themselves was a character. In fact, to kick off the event, because the tragedies were performed during the City Dionysia, a great festival and major state occasion drawing people from all over Greece, light itself was honored in the form of the god Dionysus. But fifth-century tragic drama was not a religious ritual, even though rituals were enacted preceding the festival plays, blood sprinkled and wine spilled on the stage marking out a sacred space. Dionysus was known as "*Dios phos*," the "light of Zeus," having emerged from Zeus' thigh. So the sacred space becomes light-space because "brightness" is one of the principal attributes of Dionysus who combines both "celestial brightness and torchlight." This is why the tragedian Euripides invokes him as "most radiant light."[29]

More important, tragedies were performed outdoors in a light-space filled with natural light. Performers and spectators were exposed to the sun as it shifted during the day. The audience of fifteen thousand people or more seated on the terraced stone of the amphitheater's southeast slope beneath the Acropolis focused their attention on the stage (*orchestra*) below. This was a circular dancing floor about twenty meters in diameter flanked by broad ramps leading up to it by which the audience entered, as did the actors in their entrances and exits. Behind the stage was the *skene*, a wooden structure (about twelve meters long and four high) used as a dressing room. More important, it served as a backdrop on whose façade a palace or temple or even landscape was depicted depending on what the play required.

All was public, open to the light. This is significant because sunlight in the public arena carried a powerful moral charge of honor and life, an inheritance from Homer. The worst fate in death for a Greek, dramatized in the tragedies of Aeschylus and Sophocles, was to bid farewell to the sunlight. The love of light according to their plays is as great as the pain in leaving it, because light is life itself. This is at the heart of Greek experience. Furthermore, the light, *phos*, is understood as pure and holy, thus those who violate the moral laws are unfit to look on the sun or to appear in public under the sun. Dishonored, they cannot "come out in the light," as a character in one of Sophocles' plays says. If they do so, the act not only violated the moral law but the "holy light" itself.[30] And it is this unforgiving light under which the tragedies were performed annually. In late March in the morning, specifically, the sun would hang on the eastern sky and cast a shadow across the area in front of the *skene*. This light bathes the stage and spectators throughout the festival. They are made conscious of light because the playwright calls attention to its presence.

Intimately familiar with the character of sunlight, the dramatists exploit it. They use natural light and the passage of the sun for realistic effects, as Aeschylus does in his tragedy *Agamemnon*. The action begins at dawn with a watchman seeing a beacon in the distance. But it could be the rising sun, the real sun that the audience sees. And though the beacon signifies victory over Troy, it also means bloodshed and sacrilege, so the audience is left in a state of uneasy ambiguity. The play is saturated with such ambiguity. Later in the play the sun becomes a false beacon. And the uneasiness that this generates is precisely what the playwright wants his audience to share, the real light and shadow of the theater informing the audience's view of the characters and their actions.[31]

For the ambiguous light of the dramatic world, where the contraries of light and darkness are in turn unstable, questions the real world, the real sun. Natural light, then, becomes a character on stage and off. The tragedians employ this character to mediate the mythical world of the tragedy and the real world of the audience. For light among the gods is the same light mingling with the citizens. As a result, the audience experiences the action of the play as intersecting their own world. Light also mediates the heroic past and the Athenian present. In doing so, it places the audience to some degree inside and outside time and space, prompting an ironic, critical perception of their own time and place. Although Agamemnon in Aeschylus' play enters escorted by the morning sun, for example, the audience sees the irony in this. They know from Clytemnestra's earlier testimony of his cruelty that he is blind to her situation. He stands in a light so bright he can't see. So this character, the "actor" light, is unpredictable both on stage and off. The same sun, which can be affronted by human action, "sees" the characters on stage and the spectators off stage. It is this character who ensures that the drama's ambiguities seep out into the real world. In so doing, of course, the audience sees and shares in a new experience of light's character as ambiguous and unpredictable mediator.

We can be sure that this new character of light, a Dionysiac eraser of boundaries, who bridges the mythical and natural worlds, past and present, had profound emotional and intellectual impact on the audience. He is elusive, always on the move as the tragedies proceeded with the sun's progress through the sky. The "communal meditation" of the festival, which enabled citizens from all corners of Greece to reflect on and work through their fears and sufferings, was inseparable from the space of sunlight in which it was played out.[32] Performed in the open under a relentless Greek sun, the plays of the tragedians restored the audience to a lived experience of light. For Homer's experience of radiance had over the centuries, owing to the great cultural reverence for his work, hardened into conventional symbols. Likewise, the core experience of Thales and Parmenides had solidified to abstract counters in philosophical discourse.

But Greeks now came face to face in natural light with a character having traits counter to what had become conventions in the larger culture. And this character sank deeply into their psyche. Not the light of Homer but neither the light of Thales' or Parmenides' experience, this character was now like Dionysus on stage an object of scrutiny, challenging the prevailing order. Its radiance, unlike Homer's, expressed the sacred some times and dishonor other times, blurring the traditional distinction. And just as the Athenian experience of theater permeated social practices and communication, from the visual arts to the law courts, it became a metaphor and unconscious model of human psychology.

Aeschylus ensured this to a large degree by exploiting not only the overt Dionysian nature of full sunlight, as we've seen, but by employing its elusive light-shadow dimensions. In the tragic theater we do not leave the Greek preoccupation with light and shadow, then, or their contrary relationship. Rather, the tragedians explore them in new ways. New valuations given them in the theater lead to the discovery of a new dimension of light's character, deepening the understanding of them as contraries. Thales and Parmenides employed the contraries of light and shadow as tools to probe nature. The tragic playwrights collaborate with them as partners to probe mind and character. The tragedian in order to probe personal and social depths employs the theater's natural light as it changed throughout the day with its shifting shadows cast by the skene, and as it played up the contrast between the shadowed interior of the stage building and the daylight outside.

Aeschylus' *Oresteia* trilogy, for example, which won first prize for tragedy in 458 BCE, opens with the *Agamemnon*. A drama of "doubt, distortion and partiality," it is fittingly framed by threshold times of light, dawn in the first play (*Agamemnon*) and twilight in the last (*Eumenides*). Throughout the work Aeschylus subverts the audience's expectations, not only reversing conventional associations with light and dark more than once, but in the end rendering them uncomfortably interchangeable. In the first play light rather than being a symbol of glory and knowledge is equated with crime and ignorance. It is a "perverse light" that will be by the third play transformed into a "propitious" light. Nowhere is light's character completely good or completely bad. An "evil light" in the first play used to understand the nature of Agamemnon's crime in sacrificing, or murdering, his daughter Iphigenia becomes a beneficent light later, only to shift again by the end of the play as the contraries of light and darkness remain unresolved in a "tempered union."[33]

The fact that the characters in the play continually fail to resolve these contraries dramatizes for the audience a new experience of light, which they take as their own just as they accept fear and pity resulting from their immersion in the total experience that is the tragic theater. This is borne out by the prevalence of references to the *Oresteia* on contemporary black-figure vases. The chorus, which has a central role in Greek tragedy, bravely attempts to assert the traditional associations of light and darkness. This sets up a conflict with the other characters in *Agamemnon* who pervert these associations, reversing them or questioning their relationship. Such tension makes demands on the audience, requiring that they reflect on the ambiguous character of light even as they reflect on the ambiguous actions of the trilogy's human characters. That is, Dionysus manifests himself overtly, we remember, as the "most radiant light," eraser of boundaries, but also covertly as a light-shadow trickster.

As a result, neither its characters nor the play's spectators can resolve light's character, reduce it to a single trait, any more than they can fully comprehend the motives and actions of the characters themselves. Aeschylus introduces an experience of light to the Greeks that by its very nature, like the trickster Dionysus, cannot be pinned down. It is familiar and unfamiliar. The playwright insistently undermines their expectations. From the trilogy's opening scene of light in darkness to the closing scene of religious torches surrounding darkness, this character retains a destructive and beneficent aspect throughout. Just as the beacon light becomes for Clytemnestra in *Agamemnon* a "living person rushing to Argos" bearing the ambiguous news of her husband's triumph as well as his certain end in her murderous hands, light remains a Dionysiac boundary crosser and shape-shifter to the end.[34]

This radiant character shadows Aeschylus, and this dark unpredictable character shines. It is evil and benign. It is both an actual phenomenon, literal sunlight in the amphitheater, and an imagined creation, sometimes in the costume of metaphor or symbol. It refuses to settle on either.[35] The tragedian's experience of light, then, arises out of and fosters the new consciousness. And at a time that exalts individual experience as never before, Aeschylus is not shy about revealing his core experience of light on stage in public. Dionysiac light asserts its presence on stage and among the seats lining the slope of the amphitheater. The theater invited onlookers to reflect on a space of light as well as shifting light-shadow. By prompting this, the Luminaries of the tragic theater altered the Greek experience of light. Dionysiac light lived beyond the theater as it inhabited the consciousness of its audience. We could say that light in a sense, owing to Aeschylus, has for the first time its own "psychology." Just as a few centuries later, as we'll see in the next chapter, the Refractors will discover light's "biography."

The Law of Light and Shadow

A Luminary such as Aeschylus so dazzled his audience they no doubt paid little attention to the artist who depicted Agamemnon's palace on the *skene* behind the action. Certainly, the name Agatharcus of Samos (d. 456 BCE), who was Aeschylus' scene-painter (*skenographos*), according to the Roman architect Vitruvius writing four centuries later, was not as famous as the tragedian's. Nonetheless, Agatharcus and his contemporary Apollodorus of Athens, known as the "Shadow-Painter" (*skiagraphos*), whose paths crossed in the mid- to late-fifth century BCE, were the originators of Greek painting.[36] Together, these artists introduced a stable aspect of light's character, grounding its Dionysian ambiguity.

Rooted in his experience of the theater's light-space, Agatharcus develops a new way of representing appearances that became one of the twin foundations of Western painting—perspective. What seized his imagination, as it did the Greek geometers, was the constancy of light's behavior despite the audience's point of view, which varied according to where they were seated, down close, to one side, or up high. From this experience of light's character, he found a way to express its constancy, by which he created illusionistic space, such as Agamemnon's palace architecture. Product of his theater experience, this method grew out of his ability to imagine the individual viewer's eye as a projector of light, which the Greeks believed it was. And his scene painting gives substance to the actions of this light as they play across the stage and his scenery. He observes light's edges slicing the stage, sight-lines varying predictably, and the resulting shadow shapes. Perspective allows him, then, to express in his scenery this "geometry of light" (later to be described in the language of optics).

While the geometry of light relating viewer and scene in the theater space strikes Agatharcus, the play of light and shadow relating the individual and his act of seeing captivates Apollodorus who lays the other foundation of Western painting—chiaroscuro (Italian: bright-dark). He finds expression for the individual's experience as a receiver of light in their perceiving of objects and figures given form by it in space, being the first to model his figures in light and shade, just as Agatharcus was the first to paint illusionistic architecture. We see the result of Apollodorus' discovery in the modeling of the bodies of Herakles and Telephus in a Roman copy of a Greek original painted in the fourth century BCE. The painter gives expression to a specific relation between himself and the source of light by using highlights, glints of reflected light depicted by sudden shifts from black to white, to model the forms of the body.[37]

Both Agatharcus and Apollodorus place the seeing individual as sender and receiver of light at the center of the world in effect, as if it were a theater. Out of their new experiences of light as a creator of space by its geometry and of form by its play of light and shadow, respectively, came their urge for the first time in history to represent a "depicted light." Uniquely Greek, born of theoretic and theatric consciousness, this desire ushers the self-conscious portrayal of light as light into history, which we'll see developed in the next chapter.[38] With Dionysus, light takes center stage. The geometric constancy of the sight lines of light and its reflection off the surface of an object entrance these painters, who gave expression to their experience, respectively, in perspective and modeling.

Enriched by new experiences of light expressed in a new handling of light and space and of light and shadow, what Plutarch called their "fading out and building up" in giving form (*De Gloria Athena*), these two artists established a relationship with light that would be cherished for centuries. Modeling by contrasts of lights and darks, chiaroscuro continues to hold the imagination of Western artists into our own century, influencing photography and film as well as art. Apollodorus' student Zeuxis systemized this in what he called the "laws of light and shadow," meaning that the predictable behavior of light was made explicit in a painting. In thus depicting light the painter gave solidity to opaque objects, the parts of a body exposed to the light source highlighted, and those not exposed placed in shadow according to the lines delimited by the light's rays.[39]

Flourishing a couple of generations after Apollodorus and learning from his and his student Zeuxis' work, Apelles of Kos extended his experience of light. Encountering what Pliny the elder later called light's "splendor," he finds expression for this in his depiction of "sparkling highlights" and was considered the greatest of Greek painters. Famous for painting such realistic grapes that birds came and pecked at his picture, he became court painter to Philip of Macedon and his son Alexander the Great. Apelles documented in his work a luster so bright, Pliny reports, that it risked "offending the eye." His invention of highlights to depict this aspect of light's character introduced into Western painting's expression of the human experience of light the desire to portray it per se. And with this came "optical themes," which were seized on and developed by Roman painters, becoming uniquely important in the Renaissance, as they did not in Mayan, Chinese, or Indian painting.

But this splendor existed in active relationship with shadow. Quite possibly referring to reports of Appelles' work, Pliny some four hundred years later describes the "qualities which exist" between "splendor" (or "shine" or "highlights") and "shades" as "tension" (*tonos*).[40] Apparently, he discerned in Apelles' skilled handling the core of Apollodorus' experience of light. In his experience, the contraries of light and shadow converse and argue with one another. In continual motion, they possess a Dionysiac energy. At the same time, their tension derives from the painter's taming them to the law of light and shadow (chiaroscuro) dictated by the geometry of light (perspective).

In the tragic theater Aeschylus collaborated with the contraries of light and shadow as partners to probe the human mind and character. The painters Agatharcos and Apollodoros and his students collaborate with these contraries to probe light's character directly. Their discoveries became the possession of sculptors and architects alike. And we are their heirs, taking this aspect of light's character so much for granted that we tend to regard the paths of light from viewer to scene and vice versa, expressed in the language of perspective, and light's reflecting off objects in predictable ways, expressed in that of chiaroscuro as reality itself, as the way things are, not simply the description of one aspect of light's character as experienced by an individual. This is so only because the painters gave such powerful expression to their experience of this aspect, creating the "true illusion of light." It was immensely persuasive, because as the artist and scholar Vincent Bruno notes, for the first time the painter seems to "correlate light and dark with the observed world." Apollodorus and his progeny Zeuxis and Apelles by their encounter with light thus influenced the course of Western art.[41]

The architects, likewise, who encountered light's geometric character and its light/shadow tension express it in works not only of beauty but utility. Subject to many variables, their architecture had one constant, as the French architect and scholar Eugène Emmanuel Viollet-Le-Duc observed in the nineteenth century: its existence in brilliant light. The "perfectly transparent atmosphere" of Athens meant they could count on light and shade partnering with their forms. Particularly sensitive to conditions of atmosphere and light, working within a matrix of rays, and to differences between direct and reflected lights as the sun moved across the sky, as well as to the continual play of light and shadow, they took advantage of these variables in how they structured and ornamented their buildings. They use exclusively horizontal moldings, for example, deeply sunk in order to "obtain sharp shades like strong ink-lines in a drawing" and thus create a building banded with horizontal shadows "quieting to the eye." Structurally, they use a succession of parallel planes bounding a space, making the geometric character of light explicit. But they were most concerned with surfaces reflecting light and with the "measured recurrence of columns" modulating light and shadow.[42]

The team of Iktinos the architect and Pheidias the painter-sculptor gave supreme expression to this side of light's character. Employed by Pericles to create one of the greatest buildings in the history of the world—the Parthenon (447–432 BCE)—they used brilliant white, highly reflective marble exclusively (thirty thousand tons) to construct a geometrically intricate work. Iktinos called it an "education to Greece," echoing Pericles. And so it was, high on the Acropolis not only in its unsurpassed beauty in praise of Athena, goddess of the city, but in its embodiment of a specific encounter with light that had its birth in the earlier experience of painters. Like the scene-painter, Iktinos was acutely aware of his viewer's perspective, paying particular attention to the temple's width-to-length and column-diameter to gap-between-columns ratios by which he modulated light and shadow, as does Pheidias. Furthermore, he defines the surface of the Parthenon by the play of shadows created in the undulations of an encircling frieze of sculptured reliefs below the roofline.

Such attention to the modulation of light and shadow on the Parthenon, the most monumental expression we have of Greek consciousness, is extraordinary. The frieze is an "unending fluctuation of light and dark," which the viewer must sort out, resolving in his individual apprehension by his senses and intellect. The sculptures throughout the frieze inhabit a field defined by the tension between optical and subjective dimensions of the viewer's experience as do Iktinos' columns. These dimensions embodied in the play of light and shadow give expression to the tension between the distantly observed and emotionally immediate scenes depicted in the panels of the frieze.[43] And light and dark as well as their objective geometric character behind the observer's subjective experience become Plato's preoccupation.

PLATO'S LIGHT-SHOW

Rejecting Aeschylus' representation of human action as fiction not truth, lies in effect, Plato (c. 427–347) left the theater. But it didn't leave him. As a metaphor it provided him with a way to understand the operations of the eye and of the mind. Plato the ex-playwright could shun the tragic drama but he could not escape the amphitheater's bowl of light. In this arena of looking, the spectator's relation to light moves to the foreground. Using dramatic dialogues and riding the wave of the new theoretic consciousness to call into question all previous interpretations of experience, such as Homer's, just as Aeschylus had called all moral certainties into question, Plato interrogates light itself.

Sun-Dazzle in the Theater of the Eye

Wholly arrested by the sheer dazzle power of the strong Mediterranean sun, this core experience of light was nothing short of an "epiphany," which came to occupy the center of Plato's philosophy. He could not evade the sun's pure light. This was not the white radiance, home of the gods, that mesmerized Homer, but the puissant light revealing itself as sun, and the sun revealing itself as brilliance. A revelation, this "sun-dazzle," as I'll call it, insisted Plato come to terms with it. Hence the inescapable sun was not about to be dismissed as a god, Helios, or as a "red-hot stone," which Anaxagoras taught.[44]

For Plato light is an effluent flowing from the sun that makes vision possible and the eye "sunlike," as he states in the *Republic* (VI, 508b). Sight is produced when the light from the eye goes forth and makes contact with "like" light from an object. In this event light "coalesces" with light "to make up a single homogeneous body" (*Timaeus*, 45c), one stretching between the eye and the visible object, which acts as an intermediate body between them. And when the rays of light from the object encounter this body, motion is transmitted to the viewer and produces the "sensation we call 'seeing'" Plato concludes (45d). Thus the rays of light from the stage carrying tragic action encounter the mediating light in the amphitheater constituted by sunlight and the rays of light from the spectator's eye, ultimately producing emotions of pity and terror—tragic drama.[45]

Plato's core experience of light, however, entailed a paradox of seeing. Sun-dazzle by its beauty and power demands his looking; yet looking, he is blinded. Like Parmenides he found himself in an uncomfortable contradiction from which he had to extricate himself. How can he be enthralled by the sun's variable beauty so completely yet find himself blinded by it? To resolve this contradiction, he turned to philosophy and embraced the conviction that behind this apparent chaos there lies order, an immutable cosmic order. He must find something permanent underlying the sun's appearance, which was on the face of it contradictory. In short, he could not accept on a personal or political level that this world of contradiction was reality. The state of flux between these encounters, of "becoming," could not be the real world. He needed a way out of this dilemma, and "Father Parmenides," as he called him, pointed the way. Plato found solace, as we would expect, in the father's proof of an unchanging reality, a world of "being" that underlies the flux we experience. But how to appropriate this reality, this "Heaven above heaven" for himself?

Sun-dazzle offered a solution. The sun is uniquely, he realized, "becoming" *and* "being," the difference between a particular triangle and the idea of triangle. Interpreting his epiphany of light, he convinced himself that the sun is "becoming" in its emanating the light we perceive, and "being" in its occupying an eternal realm of Ideas. We cannot look directly at the sun because it is ultimate Being (i.e., true reality). But we can see its manifestation, its reflections in the world of nature, because earthly light is Becoming.

Yet Plato yearns to see the Sun of Being itself. Although the sun belongs to the world of sense and change, it is at the same time eternal and self-sufficient. The sun remains independent of the world, though the earth depends on its light. In sun-dazzle where Being and Becoming have their unity he recognizes a character more complex than the Dionysiac light of Aeschylus and beyond what Pindar could even dream of despite his odes that "drain dry the Greek vocabulary of words for light and bright, shine and shimmer, glitter and glister, ray and radiance, flame and flare and flash, gleam and glow, burn and blaze." For Plato's is the fundamentally paradoxical sun, as he asserts in his *Republic*, that "not only provides visible things with the power to be seen but also with coming to be," and here he gives us the nub of the paradox, "although it is *not itself* coming to be" (509b). In his encounter with sun-dazzle, Plato discovers a new light.[46]

This character is so rich and complex that to convey adequately his experience of it he must ultimately turn to symbolic language, constructing from this core image of sun-dazzle a metaphysics of Being *and* Becoming, which is in fact an imagined metaphysics of light. Plato concludes from his epiphany, then, that light is a "basic substance of the universe," which can only be fully accessed in the Eternal World of Ideas. In his account of the sun as a symbol of the Good, reality itself, the Ideas have their unity. This is true because the Sun/Good resides in the "intelligible realm" of "understanding" and knowledge, as Socrates says in the *Republic*, while our everyday sun resides in the "visible" realm of "sight and visible things" (508bc). And we access these realms in two different ways. The particulars of our earthly realm of Becoming, the daylight world, are apprehended by sight in the theater of the eye. At the same time, the unity of Being in the heavenly realm of Ideas, the Light world, is apprehended by reason in the theater of the mind. And Plato's yearning will only be satisfied in attaining the Good. But how?[47]

Firelight in the Theater of the Mind

In Plato's quest for the Good or the true Sun (ultimate realm of Being), which is his project of self-education, he turns from a public encounter with the sun of the agora and amphitheater to a personal and complementary experience of light. By this he can progress towards the Good. But he can only succeed through what he calls in a letter to the followers of Dion, "communion and constant intercourse with the thing itself," intense meditation on the sun as a symbol of the Sun. Meaning that it plays a double part, the sun being both sensible, available to the senses, and symbolical. If he is successful, "suddenly like a light kindled from a leaping fire," the Good is "born within the soul and henceforth nourishes itself." Thus for Plato the way to certain knowledge is through the symbol, the visible sun. He can by "communion" with the visible sun he believes, make progress towards the eternal Sun, his goal—Being or the Good.[48]

To begin this journey, he turns to theater. Not the tragic theater but the theater of the mind. Sun-dazzle appeared in the theater of his eye where vision apprehends its performance. But to attain reality, the Sun/Good, he must enter the theater of the mind where reason and imagination perform. For Plato the former theater is an analog of the latter because "Vision is to the Sun as Knowledge is to the Good." Just as the sun illuminates the visible, that is, so the Good illuminates knowledge, and it appears as truth "born within the soul." As Socrates says, the Good "gives truth to the things known and the power to know to the knower (*Republic VI*, 508e).

Plato, therefore, must go beyond vision, physically seeing the sun in the theater of the eye, to knowledge, seeing the Sun/Good of "truth and reality" as they "shine resplendent" in the theater of the mind. At this point, having caught a glimpse at last of the Good itself, the soul "apprehends and knows them and appears to possess reason." Thus Plato furthers his education by progressing via right interpretation of the symbol, from the material to the mental theater. This is an eye-to-mind move he learned from "Father Parmenides" who first made light symbolic of philosophical truth, understanding that light is braided into the nature of mind.[49]

Plato's teacher Socrates describes this theater of the mind. He imagines a scene (commonly known as the Allegory of the Cave in Book VII of his *Republic*, 514a–520a), much as a scene-painter paints: people living in an "underground, cave-like dwelling" are chained in one place, being able to see only in front because they can't turn their heads. Light comes from a fire burning "far above and behind them." Between the prisoners and this fire some showmen walk on a roadway carrying artifacts, such as stone statues that cast their shadows on a wall in front of the prisoners. And they can only see these shadows, of course, not the things themselves (514ab).

Plato constructs his cave scene to probe the symbol, which is ultimately to imagine, or in his mind to comprehend, the true and unique nature of light. He reminds us of this by having it follow on the description in the *Republic* of his encounter with the sun-dazzle. In the eye-theater where this sun performs, it symbolizes the true Sun/Good. In the mind-theater, the relation of this symbol to reality in our world plays out, which Plato dramatizes by his cave's staging. That is, the cave is analogous to our familiar world of Becoming available to the senses, all that is visible cut off from the "real" world outside, the world of Being, all that is intelligible. The sun as a material body is to the "shadow" forms in the cave, then, as the sun-dazzle of the Sun is to the Good (Being). Both sun and shadow inhabit the world of Becoming. The Sun[-dazzle]/Good exists on a transcendent level in the world of Being.

Plato recognized that the phenomenon of light is unique as the only possible mediator of these worlds, just as light mediates the eye and the object in the theater of the eye to make vision possible. Light's unique character allows this because, as Plato knew, what reveals an object cannot be that object. That is, we see light illuminate a stone, but it is not itself stony. In Plato's terms, Being does not participate in Becoming. So sun-dazzle partakes of Reality but not of the objects in our world it illuminates, which are mere shadows of Reality. Sun-dazzle, therefore, is and is not the sun.

By his imagined scene in the theater of the mind, then, Plato achieves and expresses a true understanding of the character revealed in his epiphany of light. It is a mediator, a natural symbol of the Good. Thus by contemplating sun-dazzle as a symbol through "communion and constant intercourse," we "learn to rise up out of becoming and grasp being" (*Republic VII*, 525b), and our soul may move closer to the Good. The fire's "leaping" in the cave, continually moving in nourishing itself, is a figure of the intellect in the theater of the mind being educated to receive the vision of the Good. Thus in his cave scene with its "leaping" fire, Plato encompasses all of Being in "terms of light."[50] The "organ of knowledge" with the soul, as Plato says, must be "turned around from the world of becoming," like "scene-shifting" in the theater, until it is able to "endure the contemplation" of the "brightest region of being," which is the "good" (*Republic VII*, 518c, trans. Hamilton and Cairns).

As Plato dramatizes here, symbol is the key to the education of our soul on earth's stage because it includes in the objective sun-dazzle experience the goal of a soul's education as well as the means to this goal. Though only obtainable on earth by degrees in the subjective experience of "light kindled from a leaping fire," this is also when the Good is "born within the soul," the beginning of our true education. The educational journey ahead simply means turning from the natural sun to the manifestation of the Sun/Good, the sun-dazzle. Or as Plato puts it, "turning the soul around, away from becoming and towards truth and being" (*Republic VII*, 525c). The soul must turn from "what is mixed with obscurity" (*Republic VI*, 508d) to the pure Light.

Accordingly, Plato measures progress towards "truth and being" by the soul's mix, its ratio of light and darkness, which indicates relative degrees of knowledge and ignorance. Like Thales, Plato employs light and shadow as a tool, then, but a very different one. He uses it to measure our soul's progress on earth. This conception of light and darkness reveals Plato's conviction that a "natural" connection exists between true knowledge and light or Being itself. This is new. He is the first of the Greeks to posit such a connection. "Being as light" is true in and of itself, that is, it is not dependent on its contrast to untruth. Sun-dazzle is true being, the ground of reality and the goal of education; total darkness is simply non-being. The mingling of light and darkness is becoming, Plato understands, which is the human condition. Because as mortals, seeing the sun directly leads to blindness; we need these mingled intermediate stages in order to see it fully.[51]

This necessity is why the drama of truth in the cave is not a cosmic struggle between light and darkness but simply a human struggle to further one's education, to come into knowledge of the truth, light being both the aim and means of education. Light's unique status in the world, making objects visible and yet not sharing their nature, equips it to play this role. Owing to this unique character, light possesses the power to unite the material and the immaterial. And this points up the key difference between light and darkness in Plato's experience. Unlike light, darkness is impotent to unite the material and immaterial realms, the world of becoming and the world of Being. Thus in our world, light and darkness are contraries as they are for Parmenides, independent yet having a necessary relation. But in the world of Being they resolve to light, the Sun/Good.

Here we have in light's true character the answer to a puzzle: Plato condemns the shadow-painter's chiaroscuro but not the scene-painter's use of perspective. Apollodorus mingled lights and darks to convince us of an object's three-dimensionality. His student Zeuxis took this hint and invented a chiaroscuro that relied mainly on shading, which assumed great importance in the painting of Plato's time. In short, the shadow-painters worked in the cave of shadows cast by artificial light. The flat unbroken areas of light and color in traditional painting were now being broken up and adulterated by the artist's technique of shading in an effort to persuade us of an object's "reality."[52] But for Plato, of course, mere representation was artifice, not reality, simply a poor imitation, just as was Aeschylus' representation of human action. Thus Plato condemns in the *Republic* the "corrupting influences" of "illusionistic painting" like Zeuxis' and bans its artists from the ideal republic.

Such a technique was in stark contrast to the scene-painter like Agatharcus who recognized the geometric aspect of natural light in his experiments in perspective, suggesting its transcendent character. While Zeuxis was in the business of deceiving his viewers, taking advantage of our "natural weakness" that "shadow-painting plays upon, just like magic, and conjuring too," as Socrates charges, Agatharcus' business was "rational calculations" of the behavior of light. In Plato's view scene-painting did not deceive but hewed to what he imagined to be the mathematical underpinning of our world, the reality of abstract geometry in the realm of Ideas behind its apparent variation and change. By this Agatharcus enters the theater of the eye and the mind, adopting their terms of light.[53]

THE BONES OF LIGHT—EUCLID

Over the door of Plato's Academy, the world's first university, which endured for over nine hundred years, he posted a requirement for acceptance: "Let no one ignorant of geometry enter here." A puzzling demand from a philosopher perhaps until we learn that in his *Republic* he likens the educational journey of the soul towards the Sun/Good, via the sun-dazzle, to movement up a vertical line, indicator of the ratio of light and dark in the soul, which he describes with some geometric exactness.

Ever since Homer observed the "rosy-fingered dawn" and countless Greeks after him wondered at crepuscular rays arrowing through cloud gaps after a storm, it was clear that there was more to light than its apparent face. Sun-dazzle above, yes, but sun-puzzle below, what light could it be that brought us sight of heavenly bodies such as Venus before they were due to appear above the horizon? Or in very hot weather what light makes boats at great distances appear when they normally don't?[54] Though Plato did not address such questions, which involved the bending of light by the atmosphere (refraction), his knowledge that light had a geometrical aspect, which Agatharcus exploited in the theater, suggested an approach to solving them. The solution must wait for Archimedes (287–212 BCE), but Euclid (325–265 BCE), who probably studied in Plato's Academy, may have taken his hint in using the line of geometry as a sign.

Euclid married the abstract, a symbolic line, and the concrete, a visible ray. A stunning imaginative achievement, this unites Euclid's commitment to geometry and his experience no doubt of arrow-light. We don't know whether this familiar experience of light revealed as rays by dust motes in a room or by water particles in the atmosphere (i.e., crepuscular rays) came first, but light's revealing its straight-line character and its direction of propagation preoccupied him.[55] Euclid believed that these behaviors revealed more than aspects of light's character. He realized that such behavior exposed the very bones of light's body. The ray indicates an inner structure, which can be described by geometry. Likewise, vision is for him a geometric shape, a cone with its apex at the eye, which emits beams of light, and its extension sweeping the world as we move our head back and forth and side to side. As he postulates in his *Optics*, "suppose that the straight lines which emanate from the eye are propagated divergently." Hence he makes light in the theater of the eye an object of investigation for the first time in history.

Using diagrams, Euclid explores what he imagines as the inner character or "Soul" of light. These constitute a language enabling him not only to interrogate the structure of light by means of geometry and reason but to express his pleasure at the beauty of the realm of Ideas, as Plato believed, that the geometry of light gives him a glimpse of. And in turn, diagrams embody his discoveries about one aspect of light's fundamental character. For Euclid optics with its geometric proofs, which derived from an existential experience of essential aspects of light's character, was no doubt an aesthetic one; logic is beauty, beauty logic. As "pictorial metaphor" diagrams fuse the logical structure of his proofs and the beauty of their source in the realm of Ideas. The painters report on their encounter with light, but Euclid, like Plato, reports on his interrogation of light.[56]

In doing so, they establish divergent interpretive traditions of the experience of light. Euclid and Plato find language for experiences we all share of its sensual character: sun-arrows and sun-dazzle, respectively. They both invested this complex character, both appearance and substance, with their innermost desires. But whereas Plato saw in his experience confirmation of an order behind the world's disorder, Euclid saw affirmation of a logical order within this world. Hence, while Plato's sun-dazzle led beyond appearance into metaphysics, Euclid's arrow-light led into physics, giving him a glimpse of the bones of light's body.[57] Clearly, the phenomenon of light, its sun-dazzle and crepuscular rays, respectively, remained at the core of their experience, as it had in earlier Greek experience and before, but these perceptions in turn were transformed by their interpretive filters of Being and geometry. And these revealed new dimensions of light's character.

Looking back on the Greek Classical Age, we see that its Luminaries encountered a more complex character than did the shamans or seers of prehistoric times or even the priests of Classical Egypt. Understandably so, for the theoretic consciousness of Classical Greece, a product of cultural evolution, is overlaid on earlier mythic consciousness, gift of our physical evolution. Story-telling now invites telling about stories, reflecting on them as subjects. Similarly, light now begs thinking about light as light. And we see in this age that the human relationship with light becomes for the first time aware of itself, the Greeks giving light a "psychology." It constitutes for them the relationship of mind and world. As a result, the character of light that dominates the age reveals more dimensions of itself.

But its primary relation in the age is, of course, to shadow. This is a contrary one, as I've suggested, and central to the Greek experience. Light outlined forms the axis of the culture, as expressed in the works of Thales and Parmenides as well as of the painters and architects. Moreover, light and shadow involve the Luminaries in a new relation to light. In the hands of Thales and Parmenides, light-and-shadow contraries become a tool of discovery. As also for Plato, who used the contraries as a tool to measure the soul's progress towards the Sun-[dazzle]/Good, made possible by having resolved in his sun-dazzle epiphany the paradox of Being and Becoming.

And performing in the theater's arena as well as on its stage, light enters into a relation with its spectators. The theater invited Athenians to reflect in a space of light that concentrated the brilliant Mediterranean sunlight and placed the seeing individual at its center. Here light shows its wild side. Aeschylus presents an ambiguous and unpredictable Dionysiac character, expanding the Greek imagination of light-and-shadow contraries. The shape-shifter character of light formed the core of Aeschylus' experience, which he brought on stage for examination. By this the Luminaries altered the Greek consciousness and with it the Greek experience of light. The tragedians' unpredictable character, however, was grounded as well. Plato imagined light to be the basic substance of the universe on which foundation he built a metaphysics. Light was not only a shape-shifter, then, but more important, the fundamental constituent of reality itself.

In the same Dionysian theater, Agatharcus, in creating his painted theater scenes, and Apollodorus, in exploiting lights and shades, also discovered that the character of light had a predictable aspect. And this allowed them to create the space of the theater itself as well as the forms within. Aspects of light's character in their experience could be and were later codified, respectively, in the rules of linear perspective and chiaroscuro (inevitably becoming clichés needing to be restored by image from experience, but that's a story for later). And Euclid, having gone beyond the artists and bared light's very bones, found a language to express this experience. Making an imaginative leap, he married concrete rays and abstract lines, introducing a new language of light: optics. And what he began, the Renaissance completed in separating the experience of light from the light as object, a subject of investigation.

In short, the Luminaries encountered a light of great complexity. Examining its character, these sages discovered new dimensions in light's relations with its source, the sun, and with its contrary, shadow, as well as in light's relation to them. The Refractors of the next chapter inherit these discoveries in the theater of the eye and of the mind. But experiencing light's various acts within them and in the world at large, they probe it further to imagine light's history, eventually developing a "biography" of light.

NOTES

1. "Writes on the ground," Serres, "Gnomen," 76, 79, 86–87. Divorce of the sun from the gods, Patricia O'Grady, *Thales of Miletus*, 5, 148, 150.

2. Bellah, *Religion in Human Evolution*, 271, 273, 277. Robert Bellah employs in his account of the dawning of the "Axial Age," *Religion and Human Evolution*, Merlin Donald's analysis of the four stages of human evolution: (1) *episodic*; (2) *mimetic*; (3) *mythic*, which he describes as "a unified, collectively held system of explanatory and regulatory metaphors," where the mind has "expanded its reach beyond the episodic perception of events, beyond the mimetic reconstruction of episodes, to a comprehensive modeling of the entire human universe" (272); and (4) *theoretic* culture in which, in Donald's words, "the human mind began to reflect on the contents of its own representations, to modify and refine them" (quoted by Bellah, *Religion and Human Evolution*, 364). Greek culture was the first to give rise to full-blown theoretic culture accompanied by the "efficient external memory system provided by a fully alphabetic writing system" (*Religion and Human Evolution*, 273). And this influenced, as we'll see, the experience and imagination of light. This now arose in Greece, that is, the "ability to think analytically rather than narratively, to construct theories that can be criticized logically and empirically" (274). But narrative or mythic culture "augmented," it did not replace, mimetic culture, just as the theoretic augmented but did not replace the mythic (364).

This last stage of evolution, the theoretic, which produced a "systematization of inquiry," differs from the previous three stages (Bellah, ed., *The Axial Age and Its Consequences*, 69). Instead of resulting from our physical evolution, it results from our cultural evolution. The first stages, that is, are hard-wired; the last is the result of hard-fought cultural struggle (71). This is due in part to the fact that each stage is layered on top of the earlier stages, none lost, so our consciousness is a laminate comprising evolutionary and cultural products of consciousness (and see Donald, "Art and Cognitive Evolution," 7–10), a layering revealed in the subsequent history of light in human experience.

3. "Transparent clearness," *A Dictionary of Greek and Roman Geography*, quoted by Smith, "Athenae," I:255; "powerful influence," C. M. Bowra, *The Greek Experience* (Cleveland, OH: Mentor, 1957), 11; "understandable to itself," Andrewes, *The Greeks*, 12; Blumenberg, "Light as a Metaphor for Truth," 33, 46. The principal way to certainty for the Greeks was through the eye. For them truth came by seeing not by hearing as it did for their Hebrew contemporaries.

4. Blumenberg, "Light as a Metaphor for Truth," 33. They conceived this quite literally. Explaining how distant light can affect us with its image, they postulated that the image was a "skin-like material effluence" from the visible body that traveled the medium of light and touched the eye. Hence the image, which is mind itself, inseparable from the physical world (Authier, "Refraction and Cartesian Forgetfulness," 317; "skin-like material effluence," Cantor, "Physical Optics," 629). The Greek experience of crystalline light was thus too comprehensive to be isolated in a deity. This accounts for the fact that although Greece had many gods, Greek religion had no god of light. Helios (sun) was a Greek sun hero not a sun god, having no cult until Roman times.

5. "Act *of* consciousness not . . . *in* consciousness," quoted by Kearney, *Poetics of Imagining*, 90, 92, 100. I use Gaston Bachelard's distinction between "image" and "metaphor." The image is imagination itself, fundamentally bound up with perception, whereas the metaphor is illustrative, derived from perception. The image, as Bachelard asserts, is the "pure product of the absolute imagination," in itself a "phenomenon of being." Metaphors come after. They give "concrete body to impressions otherwise difficult to express," but as "mere mental illustrations of perceptions and 'images proper'" do not express the "essential dynamism of being" because they bear an "accidental relationship to the being of the imagination" (quoted by Kearney, 99). Our apprehending an image means we must find a unique language to express it. This language is metaphor.

6. Brilliant, *Arts of the Ancient Greeks*, 133.

7. "Silhouette principle," Robertson, *A History of Greek Art*, I, 214. Vase painting gave way by the fourth century to free painting on wood panels and walls as it moved beyond line drawing and expanded with Apollodorus the "Shadow-Painter" of Athens' discovery of what came to be known as "chiaroscuro." Dedicated to appearances, this painting captured the public and consolidated its position in the history of art with the "naturalism" of Apelles of Kos who was revered as the greatest of Greek painters.Yet even when Apelles' four-color practice dominated in the fifth century, the imagination of light and darkness prevails, which the early vase painters had imparted to their fellow citizens, not the principle of complementary colors. The palette was restricted to just Apelles' four, yellow and red, white and black (Bruno, *Form and Color in Greek Painting*, 83), two sets of high-contrast, light-dark values.

8. Bruno, *Form and Color in Greek Painting*, 85.

9. Stewart, *Classical Greece and the Birth of Western Art*, 13.

10. Richter, *Attic Red-figured Vases*, 3. The tension between the conventional norms of society and their transgression is expressed in the vase painters' images as well. Many images of women, for example, reinforce their role in the home as cloth-makers and child-bearers. At the same time, images of courtesans abound. More important, as Mary Beard shows, images appear that undermine the stereotype and even present transgressors of the norm ("Adopting an Approach," 21–34).

11. Mertens, *How to Read Greek Vases*, 100.

12. "Light a clear path . . . black vessel" and "stretches out over . . . sea," Knox, *The Norton Book of Classical Literature*, 221 and 226.

13. Halfway around the world in China during the Zhou period, a contemporary of Homer's addresses Heaven, "O Radiance, O Light" (Bellah, *Religion in Human Evolution*, 258). *Dictionary of Untranslatables*, 576–77; Constantinidou, "The Light Imagery of Divine Manifestation in Homer," 91–92, 94.

14. Mesopotamian and Egyptian influence, Constantinidou, "The Light Imagery of Divine Manifestation in Homer," 99. Blumenberg, "Light as a Metaphor for Truth," 33. Homer's comprehension of light's radiant power to grant existence in our world is a familiar aspect of light's character from earlier times. But rather than complementing as night does day, darkness erases the light of life. Darkness is the god Erebos born directly from Chaos, whose sister and consort is night. Night's appearance can be understood in relation to the day's light. But Erebos cannot; it is the primeval, perpetual darkness defined by eternal absence of sunlight (Marinatos, "Light and Darkness and Archaic Greek Cosmography," 193, 198–99). When Homer has a warrior in the *Iliad* who knows he is about to be killed beg Zeus that it be done in the light, "Destroy us in the light, since [my death] is your pleasure," we're puzzled. Until we realize that death in the dimness of the fog which has descended on the battlefield would be a foretaste of the sorrowful afterlife in Hades, the eternal sunless world. The warrior begs for light, for life itself in the light, the substantial palpable light in which things have their very being, declaring themselves in radiant outline (Calasso, *The Marriage of Cadmus and Harmony*, 337). The warrior's plea is poignant because he knows his brief time in the light on earth is at an end. And he wants in the moment of death to prolong his time in the light.

15. "Children of Light," "ultimate reality" (Bose, *Hymns from the Vedas*, 5, 261); "flowing streams of light," "dazzling light," quoted by Doniger, *The Rig Veda*, 57 [RV, 10.58b] and 121 [RV, 8.79–80]; Eliade, *The Two and the One*, 26, 43. Radiance confronted gives rise to both the Greek and Indian traditions. But if it is the blaze of noon that pierces Homer most deeply, it is the glow of dawn that lodges in the Vedic poet. Personified as a goddess, Dawn is radiant, possessing freshness and grace. Her "body is bright, like one who has bathed" (Bose, *Hymns from the Vedas*, HV, 27). In her presence the poet, stunned, can only stammer the obvious, that she "stands visible to us." It is this glow that his personal experience of light begins and ends with. So radiance is central to both traditions. And though the epic poet and Vedic poet seize on different aspects of radiance—its blaze at noon and glow at dawn—both are struck by its existential power. They share as well the experience within a mythic consciousness of light and darkness as complements rather than contraries. The Vedic goddess of Night is the sister of Dawn and like her a bright creature full of stars just as she is of the sun. The poet asserts that "The dark day and the bright day, the two realms of space, turn by their own wisdom." Radiant and black, the sun rests each day to be freed at dawn (quoted by Doniger, *The Rig Veda*, 216,

and see 199; 116 [RV, 6.9.1]). These complementary phenomena constitute the universe, as they do for Homer, within the mythic mindset to become with the rise of the theoretic consciousness, contraries.

16. Haney, "Metaphor and the Experience of Light," 237–38.

17. "White radiance of sunlight," Tarrant, "Greek Metaphors of Light," 181. That is, until the literal radiance as a metaphor of a god's presence becomes in a personal epiphany again literal. Or until the metaphor dies and the literal and figurative merge in religion, its "meaning" reified and accepted as "real," not merely figurative. The brightness of Grecian daylight that seizes Homer is for him, as we've seen, a divine epiphany. As the people make his poetic expression their own experience of light's character, they create a literal form of these epiphanies to use in their cultic practice, such as Pallas Athena's golden lamp described in the *Odyssey*. Popular cultic practice employed elaborate marble lamps, and the gods even in Thales' time typically bore torches, which many black-figure vase paintings record. For Homer's countrymen the lamp, a human creation, was the conduit of divine radiance, which was experienced by the common man as Athena's gift of brightness (Parisinou, *The Light of the Gods*, 7, 163–64). And this was the thrust of popular initiation rites and hymns of the age as well, established by the shaman Orpheus. However, this gift is attributed in these not to Athena but to the god Phanes who figures in an alternative cosmic myth from that of the *Iliad* and *Odyssey* (*Oxford Encyclopedia of Ancient Greece and Rome*, 5:121). In the Orphic story of creation, the first-born from an egg made by Cronus is "Phanes," called "dazzling light," who is the bisexual, invisible creator of all, radiance itself. Existing before the sun that s/he created, he both makes visible by her blaze of light and is light.

18. Similarly, the Upanishad writer knows pure light is not simply literal or metaphoric but a symbolic tool he can use to probe his inner self. Leaving the ancient epic and Vedic poets, the Indian experience continues to parallel the Greek. In Thales' age, priestly commentaries on the Vedas called Upanishads reveal a theoretic consciousness with a new apprehension of light's character, just as we saw in Greece at the beginning of this chapter. The mythic account of creation begins traditionally, for example, "There was neither non-existence nor existence then. . . . What stirred?" But contrary to what we might expect, the answer to this and to each of a whole series of questions in the Upanishads is the question, "Who really knows?" (Doniger, *The Rig Veda*, 25). This reply suggests that the poet stands to some degree outside his received tradition, for he tacitly admits that it doesn't contain all the answers. Perhaps its model of the universe and of light needs adjusting. Among the speculations of the Upanishads, we see the theoretic consciousness clearly emerging from the mythic soil and consequent new valuation placed on radiance (Bellah, *Religion in Human Evolution*, 513). But in the Upanishads its valuation is radically different from Thales'. He looked outward to the natural world, reflecting on it and giving rise to natural philosophy. In contrast, the Luminaries of the Upanishads continue a long tradition from the Vedic Age and look inward to human nature, speculating on it and giving rise to a systematic psychology. While the sun's rays for Thales provide him with the core image of light outlined, rays in the Upanishads "start from yonder sun and enter into" the "arteries" of the speculating writer, and "they start from these arteries and enter into yonder sun" providing him with the primary image of a light-linked inner and the outer world. The Upanishads name this sun "Brahman," who is "self-luminous and formless" as well as "uncreated and existing both within and without." He is not so much a god, however, as the philosophical ground of being, the "real," the true immortal self. Hence the real is the "Bright," the "radiant" (Nikilananda, *The Upanishads*, 113, 130, 304, 352).

We experience light as immaterial, as if it pervaded our body, and at the same time as material, an object. In this paradox lies a fundamentally contrary relationship. The Western sage like Thales separates his experiencing inner self from the outer phenomenon, the light at noon; the Eastern sage, such as the Upanishad authors, collapses these internal and external experiences, the glow at dawn within consciousness merging with the phenomenon of dawn's glow outside consciousness (Eliade, *The Two and the One*, 26; "Spirit, Light, and Seed," 4). This is why the questions of the sage in India led to metaphysics and religion—Hinduism and Buddhism—whereas in Greece they led to metaphysics and science. In both cases the experiential, an individual's comprehension of the dawn's light in India or of the noon's in Greece, contributed to a momentous cultural shift in the apprehension of light's character.

19. Chidester, *Word and Light*, 129. David Chidester traces a "path" from the perceived image through "conceptual thought" to "symbolic patterns of articulation" and back to their "perceptual ground." The original image is thus always "implicated in symbolic discourse through metaphors," always "grounded in perception." It is "predetermined by physiology, determined by culturally constructed associations, and overdetermined" by the "resistance of the living body," the core image, to "symbolic reduction." In this dialectic, "perceptual metaphors" are "generated by the body" (*Word and Light*, 29, 129) by our sensory apparatus. And the "metaphorics" of light put up their own resistance to a single meaning. Metaphor by its nature precludes any absolute vantage point that could be declared the Truth. We are all "privileged" observers, all at the center of a metaphoric circle. In other words, it's metaphors all the way down. The danger is that moving from the physical to the symbolic or metaphysical buries this fact, so it is forgotten and a privileged position taken, which declares that "real" knowledge has been attained. But there is no stance that allows us to get outside metaphor. There is no non-metaphorical perspective, that is, from which to view the "real" or the "true" (Corrigan, "Light and Metaphor in Plotinus and St. Thomas Aquinas," 188–89).

20. "Wind-swift motion of brain," quoted by Stewart, *Classical Greece and the Birth of Western Art*, 111–12; Bellah, *Religion in Human Evolution*, 378–79.

21. "Shocked and overwhelmed," Popper, *The World of Parmenides*, 101.

22. Burnet, *Early Greek Philosophy*, 172.

23. Though I differ in my conclusion that he redeems observation, my interpretation of Parmenides' journey relies on Karl Popper's argument (*The World of Parmenides*, 72, 80, 86–88, 90, 100–01, 125–26). In the second part of the poem, "The Way of Opinion," Parmenides allows his readers to appreciate the shadow play of light and night because they now know that the only real thing is the "material Moon," which is independent of its illumination. He frees himself from perplexity, then, because his logical method has revealed to him the true invisible reality of the moon. Understandably, he maintains an ambivalent relationship with the world of experience. Speaking in the words of the goddess, he rejects it yet includes it, even as he has the goddess warn that her words are deceitful. As a result, ironically, Parmenides laid the foundation for trusting the senses and a thorough-going materialism. And this resolves the puzzle of why he expresses so poetically and forcefully in "The Way of Opinion" a "view he believed to be false" (Burnet, *Early Greek Philosophy*, 182, 184). For a summary of interpretive approaches, see Palmer.

24. Burnet, *Early Greek Philosophy*, 172. There is as much debate over the exact direction of the route as there is about the precise meaning of the allegory. Arnold Hermann admits that his speculation of its being an actual route is a "remote possibility" (Hermann, *To Think Like God*, 164–65). Yet Parmenides' personal urgency and the existential crises that generated the poem argue for true allegory, which links the metaphorical with the literal.

25. "Light and night," Hermann, *To Think Like God*, 177; Popper, *The World of Parmenides*, 72.

26. "All things . . . light and night," Burnet, *Early Greek Philosophy*, 176. For a counterargument, light as "pure Being," changeless, and darkness as the "non-being of light," where Parmenides is not an empiricist but a mystic, see Vlastos, *Studies in Greek Philosophy I*, 158–62. For light and night in Parmenides as complementary, see Fränkel, *Early Greek Poetry and Philosophy*, 360–63. Tarrant maintains that in the allegorical opening to his poem Parmenides regards knowledge as coming into "brightness," and ignorance as remaining in darkness; light and night being in opposition ("Greek Metaphors of Light," 183). But the opposition is between the light-shadow (night) appearances and the moon's reality.

27. Lattimore, "From Classical to Hellenistic Art," 459. In Sophocles' tragedy *Antigone* (produced around 442 BCE) he gives to the chorus an "Ode to Man," which doesn't mention the gods until the end, an afterthought. "Wonders are many on earth," the chorus sings, "and the greatest of these / Is man" (quoted by Stewart, *Classical Greece and the Birth of Western Art*, 111–12). Human dignity and liberty rather than God and king occupy the center of society, allowing democracy to radiate from it (Brilliant, *Arts of the Ancient Greeks*, 128). And for the first time sculptors create individuals, a real athlete in action, not an idealized athlete frozen in time, participants in the stream of human experience, defined by "flickering shadows" and subtle transitions as the light plays over their surfaces (Pollitt, *Art and Experience in Classical*

Greece, 69). The statutes of luminous human bodies by contemporary sculptors make the same point. Previously, luminousness was reserved for gods or the god-king (e.g., *Hydria* by the Berlin painter, Pollitt, *Art and Experience in Classical Greece*, figure 8, 21).Yet Greek religion harbored a paradox. On the one hand, Protagoras confesses, "Regarding the Gods, I have no means of knowing whether they exist or do not exist." But on the other, religion pervaded every aspect of Greek life. "Common sacrifices" at all levels of Greek society bound all groups together and served to define what a Greek was. This is why when Protagoras' contemporary Anaxagoras (500–428 BCE) denied that the sun is a god, declaring that it was merely a "mass of hot iron," he caused an uproar in Athens and escaped death only because its ruler Pericles intervened (Collis, *The World of Light*, 123).

28. Bellah, *Religion in Human Evolution*, 354–57, 600; Stewart, *Classical Greece and the Birth of Western Art*, 6, 8.

29. Taplin, *Greek Tragedy in Action*, 161, 165; "light of Zeus," Kerenyi, *Dionysos*, 74–75, 279; "celestial brightness and torchlight," Parisinou, *The Light of the Gods*, 88–89; "most radiant light," Euripides, *The Bacchae*, l. 60. The "double light of Dionysos (divine radiance and torchlight) is a direct reflection of his double nature, the divine and the human, as he admits himself" (l. 53) (Parisinou, 88–89). Pindar remarks on the "pure light of high summer," which is Dionysus or created by him.

30. "Come out in the light," Bernidaki-Aldow, *Blindness in a Culture of Light*, 11, 18–24.

31. Starkey, "Play in the Sunshine."

32. "Communal meditation," Armstrong, *The Great Transformation*, 268.

33. "Doubt, distortion and partiality," quoted by Taplin, *Greek Tragedy in Action*, 141; "propitious," "evil light," "tempered union," Paul, "The Propitious and the Perverse," i, 0, 3, and see 11. The black-figure vase artists were enamored of this "union," Taplin, *Pots and Plays*, 48.

34. Paul, "The Propitious and the Perverse" 23; Russo, 1; "living person rushing to Argos," Russo, 34. Because, as N. M. Russo says, the *Oresteia* is "replete with imagery of light and darkness. But it is often difficult to determine whether or not a specific image should be considered as such." Many images are obvious, but "others are more ambiguous." This is to be expected in a trilogy whose "recurring themes" are "ambiguity, irony, and somber foreshadowing, especially through the interplay of light and darkness." He goes on to say that the ambiguities are "solved on a divine level" in the last play of the trilogy where the "ominous image of light is transformed into brilliant torches of true joy," yet these appear in the twilight and the joy is confined to Agamemnon's house (Russo, "The Imagery of Light and Darkness in the Oresteia," 3, 7, 155, 208, 210).The ambiguities continue to haunt any conclusion the audience might tentatively arrive at. John Peradotto agrees with Russo, maintaining that the "wholesome light" and the "evil light" that "mingle ambiguously" in the middle play, harmonize in the last play, the *Eumenides* (Peradotto, "Some Patterns of Nature Imagery in the *Oresteia*," 379). As Judith Paul points out, however, the character of light throughout the trilogy is never completely "wholesome" or "evil." This is critical to a trilogy that shines a ruthlessly bright light on the "dark underground of human existence" (K. V. Fritz, quoted by Blumenberg, "Light as a Metaphor for Truth," 33). These plays analyze with the aid of the Dionysiac character of light human motives and rationalizations in astonishingly searching and sustained ways. Light, therefore, has an "ambivalent function" in the trilogy (Vickers, *Towards Greek Tragedy*, 347, 358). This moves the audience to reflect on the operation and outcome of human choice, which appears to be Aeschylus' intention.

35. Peradotto, "Some Patterns of Nature Imagery in the *Oresteia*," 388. Aeschylus is not alone. Sophocles, however, in his *Oedipus* and *Ajax* is more intent than Aeschylus in finding a resolution to the paradoxical nature of this light (Holt, "Light in Sophokles' *Trachiniai*," 205–07, 212, 215). And although Euripides falls back on conventional associations in his *Alcestis* (Bernidaki-Aldous, *Blindness in a Culture of Light*, 12, 15, 17, 147, 149), his *Medea* confronts the sun's ambivalent position in the theater with its metaphorical as well as physical brightness (Syropoulos, "S-light Anomaly: Dark Brightness in Euripides' *Medea*," 82–84). Medea is as paradoxical as Aeschylus' Dionysiac light.

36. Summers, *Vision, Reflection, and Desire*, 17, 23.

37. Although no examples of Agatharcus' work survive, his textbook on scene painting has, and we can infer his effectiveness at creating three-dimensional spaces from later Roman wall paintings influenced by this Greek discovery, such as Herakles and Telephus, a wall painting from Herculaneum (National Museum, Naples). Robertson, *A History of Greek Art*, I: 414; "geometry of light," Summers, *Vision, Reflection, and Desire*, 17–18; Gombrich, "The Heritage of Apelles," 3.

38. "Depicted light," Summers, *Vision, Reflection, and Desire*, 18. This had not been a feature of any past painting, neither Egyptian nor earlier Greek. Nor was it a feature of painting by Agatharcus and Apollodorus' contemporaries in India (e.g., the frescoes at Ajanta), China, or Mexico (Mayan painting) (Summers, 18). The introduction of "depicted light" completed the shift in Greek painting from an idealistic or "conceptual" style to a naturalistic or "perceptual" one, which became central to Western painting (Stewart, *Classical Greece and the Birth of Western Art*, 206). That is, they developed a language to express their sense of rounded form as distinct from the flat silhouette. This was the "three-tone code for 'modeling' in light and shade" (i.e., a neutral tone and two modifications toward light and darkness), which underlies all later developments in Western art (Gombrich, *Art and Illusion*, 40–41).

39. "Fading out and building up," Pollitt, *The Art of Greece 1400-31 B.C.: Sources and Documents*, 112.

40. "Sparkling highlights," quoted by Summers, *Vision, Reflection, and Desire*, 41, and see 16, 18; "offending the eye," quoted by Gombrich, "The Heritage of Apelles," 17. Greatly enamored by this particular aspect of light's character, "splendor," Apelles invented a thin black varnish, almost invisible to the eye, which when used as a coating brought out the brightness of his painting (Pliny, *Natural History*, XXXV, 79–80, quoted by Pollitt, *The Art of Greece 1400-31 B.C.*, 168). This was so effective that some onlookers claimed they could only look at Apelles' paintings through dark glasses (Gombrich, 17). And "qualities which exist," Pliny, *Natural History*, XXXV, 29–30, quoted by Pollitt, *The Art of Greece 1400-31 B.C.: Sources and Documents*, 228).

41. Zeuxis' invention of a "painterly system" of chiaroscuro, as Quintilian in the first century CE notes, allowed the "true illusion of light." And Parrhasius, a contemporary of Zeuxis, creates space that we feel we could walk into, joining the depicted figures. This is owing to the sense that the "space is filled with light falling from a single source." This light and dark is for the first time "correlated with the real world around us." Gone is the artificial picture plane that surfaces abstract geometric design. And although the human figure has not changed the "treatment of space and light around," it is "far different." We now have a figure developed, according to Quintilian, by Zeuxis "set within an atmosphere of visually rational, psychologically convincing light and shadow." Zeuxis, to the annoyance of Plato, delighted in illusionism, a threatening naturalism. He went so far as to carry "chiaroscuro to the point where its special laws could be . . . systematized" and exploited (Bruno, *Form and Color in Greek Painting*, 38–40, 43). So successful were Agatharcos and Apollodoros in revealing the new face of the light they encountered that painting was elevated in status, being esteemed the equal of sculpture and architecture for the first time in Greek culture.

42. Summers, *Vision, Reflection, and Desire*, 155; "perfectly transparent atmosphere," Viollet-Le-Duc, *Lectures on Architecture*, II, 241, and see 177; "measured recurrence of columns," Carpenter, "The Aesthetics of Greek Architecture, " I:124–26).

43. "Education to Greece," quoted Stewart, *Classical Greece and the Birth of Western Art*, 136. Even in the last stages Iktinos made fine adjustments to the building's geometry. He ordered that the columns lean inward a bit, deviating from the older conceptual way of imagining a building and, instead, insisting on a perceptual approach. In this he expresses a tension that we experience as disconnect between what we expect to see and what we actually see (Pollitt, *Art and Experience in Classical Greece*, 76).The effect of a sculpture such as Iris, goddess of the rainbow on the west pediment, for example, resides almost exclusively in the vibrant play of "irregular patterns of light and shade" over its surface, "unending fluctuation of light and dark" (Pollitt, *Art and Experience in Classical Greece*, 83, 90). Pheidias' control of the hard Mediterranean light in this drama is masterly. And he was not alone. While the Greek sculptors knew that stone is the "habitat of soft light," an "animating light" essential to their statues, they also knew the power of the Mediterranean sun (Milleker, *Light on Stone*, 11).

They had to tame this light and the resulting tension served to animate their works. Ingeniously, some sculptors created separate fields of light to control it. In the stunning "Aspasia" type of sculpture we know from a Roman copy of an original done about 460 BCE, for example, the woman's woolen cloak (*himation*) covering her entire body is rendered in a "flat, wide plane" from her shoulder to her elbow and ankle, which results in an imposing geometric shape resting on a base (Pollitt, *Art and Experience in Classical Greece*, 38–39, figure 16). But we see triangles of light produced by the fold from shoulder to opposite hip defining a separate field of light. The sculptor brings light under control by exploiting his received experience of its geometry inherited from the sculptural tradition within which he works. And by this he reveals the underlying form of the work.

44. "Epiphany," Steinberger, *The Problem With God*, 147; "red-hot stone," Notopoulos, "The Symbolism of the Sun and Light in the *Republic* of Plato," I, 171. Peter Steinberger recognizes his experience as an "epiphany," but one about what knowledge we can have of God, not about the character of light, though he recognizes the paradoxical nature of Plato's sun. Namely, the sun makes it possible to see but we cannot see the sun itself. In Steinberger's view, Plato's sun is a metaphor for how the "underlying principle or process or entity" of the universe functions.

45. All quotes from Plato are from *Plato Complete Works*, ed. John M. Cooper, unless otherwise indicated. By a fire burning in the theater of the eye, as Plato explains in his account of the creation (*Timaeus*, 45b), we perceive the fire of the sun, a notion based on Empedocles' theory of vision that "like perceives like." And sight itself is supreme among the senses because unlike the others it requires a third element in order to "see," one intermediate between the eye and its object, namely light. Without it "statements about the universe" (47a) could not be made. This is why the eyes were first of the organs to be created (45b), and the spectator's dependence on light established.Plato's star student Aristotle disagrees, saying that light is "neither fire nor body generally nor an effluence . . . but the presence of fire or something fiery in the transparent." He defines the "transparent" as that which, "though visible, is not, properly speaking, visible in itself but by reason of extrinsic color," such as air and water. These share a "certain natural attribute" which is "present also in the eternal body on high. Light is the actuality of this transparent *qua* transparent." And darkness is "absence from the transparent" of this attribute; "plainly light is the presence of it" (*On the Soul*, trans. R. D. Hicks, quoted by Cohen, *A Source Book in Greek Science*, 285–86). In short, light is not substance as it is for Plato but action, the "activity of what is transparent," as David Park says, *The Fire within the Eye*, 42 (and see Lindberg, *Theories of Vision*, 5–6).

46. Notopoulos, "The Symbolism of the Sun and Light in the *Republic* of Plato," II, 223; "drain dry the Greek vocabulary of words for light . . . burn and blaze," Gildersleeve quoted by Notopoulos, I, 165.

47. "Basic substance of the universe," Notopoulos, "The Symbolism of the Sun and Light in the *Republic* of Plato," II, 226.

48. "Suddenly like a light . . . nourishes itself," Letter VII, quoted by Notopoulos, "The Symbolism of the Sun and Light in the *Republic* of Plato," II, 240; Murphy, "The 'Simile of Light' in Plato's *Republic*," 97. Derrida asserts that the sun is a "paradigm of the sensory *and* of metaphor," which means that it is "no longer completely natural," an "*artificial* construction" (quoted by Haney, "Metaphor and the Experience of Light," 238, ital. Derrida's).

49. "Vision is to the Sun as Knowledge is to the Good," Notopoulos, "The Symbolism of the Sun and Light in the *Republic* of Plato," II, 228, and see 169–70. We catch a glimpse of this realm in the tale of Er concluding Plato's *Republic* where we hear a report from a man who has come back from the dead, having seen the other side. The soul on its journey, he says, sees a "straight column of light," something like a rainbow but "brighter and more pure." It "stretched over the whole of heaven and earth." And the "light itself . . . binds" the heavens together "like the cables girding a trireme." From their ends "hangs the spindle of Necessity, by means of which all the revolutions" of the cosmos turn (616bc). That is, pure light is a "basic substance of the universe" (Borzacchini, "Light as a Metaphor for Science," 155).

50. Blumenberg, "Light as a Metaphor for Truth," 32–33; "stage property," Notopoulos, "The Symbolism of the Sun and Light in the *Republic* of Plato," II, 237. Plato does not, however, make the fire in the cave analogous to the sun outside the cave. As James Notopoulos

argues, the "light condition in the cave was not intended to be a counterpart to the light above." The fire is a "stage property" not "a symbol of the visible world" ("The Symbolism of the Sun and Light in the *Republic* of Plato," II, 237–38). Dorothy Tarrant argues the opposite, however, in what Notopoulos considers a misunderstanding as old as Theophrastus ("Greek Metaphors of Light," 185). For "terms of light," Notopoulos, II, 235; all tied together by the "simile of Light," N. R. Murphy, "The 'Simile of Light' in Plato's *Republic*," 94.

51. "Being as light," Blumenberg, "Light as a Metaphor for Truth," 32; Notopoulos, "The Symbolism of the Sun and Light in the *Republic* of Plato," II, 230.

52. Blumenberg, "Light as a Metaphor for Truth," 33; Bruno, *Form and Color in Greek Painting*, 29. Andrew Stewart considers it "odd" that Plato does not attack the scene-painters as he does the shadow-painters with their chiaroscuro and the sculptors who "adjust the proportions of their statues in order to make them look better" (*Classical Greece and the Birth of Western Art*, 207).

53. "Corrupting . . . painting," "natural weakness . . . calculations," quoted by Stewart, *Classical Greece and the Birth of Western Art*, 206.

54. Authier, "Refraction and Cartesian Forgetfulness," 315.

55. He knew that Leucippus' *eidola* traveled in straight lines, as did Empedocles' rays. And, of course, these rays would be embraced by a geometer who would apply geometry to seeing, more precisely to the behavior of the eye in the act of seeing. His theory of vision is basically Plato's, "a bundle of visual rays issues from the eye and in some unexplained way senses the thing it strikes" (Park, *The Fire within the Eye*, 53–55). Arthur Zajonc agrees, but notes that Euclid's "mathematization" of light was the end of experiencing (*Catching the Light*, 25). But such method of description remains rooted in experience. There is no reason why his approach should preclude direct experience, only extend it. Euclid's abstracting arose from and returned to experience, testing it. Admittedly, he moves from the "lived experience" of light and makes it a "formal object of investigation" (*Catching the Light*, 26), but we can't separate the "lived experience" from the culture that shapes it and is shaped in return. As David Summers reminds us, "light is conditional" and its "representations" are "culturally specific, . . . continually shaped by human constructions" (*Real Spaces*, 467). Michel Authier succinctly captures this reciprocal constant of our experience: "The physical description of phenomena affects our perception of them and sensations are transformed by the knowledge of theory" ("Refraction and Cartesian Forgetfulness," 317).

56. "Pictorial metaphor," Snyder, "A Man Who Shaped Our World," C8; "pictorial metaphor," Kahler, "The Nature of the Symbol," 60. And see Summers, *Real Spaces*, 486–87. Agatharcus who along with Democritus and Anaxagoras made contributions to the theory of painting and to optics would have recognized this light of Euclid, being familiar with its habit of propagating in straight lines revealed by the sharpness of shadows. And ever since, Western art has developed hand in hand with optics. David Summers clarifies the relationship between painting and optics, the "*geometric* description of light," saying that optics is a "representation of vision before it is specifically pictorial representation of light" (*Vision, Reflection, and Desire in Western Painting*, 2, italics his). This is because light seen through the lens of geometry takes place in abstract space according to a set of abstract rules (Authier, "Refraction and Cartesian Forgetfulness," 320–21).Euclid was not alone in his interrogation. Halfway around the world, Euclid's contemporary in China known as the Mohist writer also probes light's anatomy. Like Euclid, this anonymous writer of parts of the anthology known as the Mozi, the main source for "Mohist" thought, filtered his experience through a commitment to geometry, seeing light-rays as lines (Needham, *Science and Civilization in China*, v. 4, part 1, 81).

57. Plato, in his concern to free this character from its *cosmic* matrix, any identification with the gods whom he found implausible, nevertheless, charged light with symbolic and ultimately metaphysical significance. In other words, Plato viewed light's appearance through the lens of Parmenides' proof of Being, whereas Euclid viewed it through the lens of previous geometers and Pythagoras' theorems. Plato may appear to move away from his core experience of light, the sun-dazzle, but it is worth remembering Michel Serres' aphorism, "metaphysics is a metaphoric physics" (Serres, *The Birth of Physics*, 43). Euclid, likewise, also moved in this abstract direction in his concern to free the character of light he experienced from its *natural* matrix,

any palpable connection with actual visible rays. By applying a persuasive, consistent system like geometry to optics, Euclid "turns certain kinds of optical illusions," secondary experiences of light such as parallel lines appearing to converge and square towers appearing to be rounded in the distance, "into the objects of a science" (Berryman, "Euclid and the Skeptic," 183). The two traditions stemming from Plato and Euclid, respectively, metaphysics and physics whose practitioners Plato calls the *Theologoi* who "speak about the gods" and the *physiologoi* who "speak about nature," separate farther in the centuries to follow as philosophy parts company with science. Though he rejected the gods as without "certain proofs" (*Timaeus* 40e), Plato counted himself among the former because he retained the One, the "god" who gives purpose to the cosmos.

Chapter Five

Acts of Light, the Refractors

The Roman Classical Age

A student of Plato, Euclid worked at the "Museum," a school founded in 300 BCE that included the famous Library of Alexandria where he could have read Aeschylus and Aristotle. The center of cultural life in the city of Alexandria, its success charts the shift of power from the Greeks of our last chapter to the Romans of this chapter, as they rose in fewer than fifty-three years (220–167 BCE) to govern most of the Mediterranean lands by the year Julius Caesar died in 44 BCE, eventually falling to a Visigoth army from the north that sacked Rome in 410 CE. During this period extending through the demise of Greco-Roman pagan culture and its eclipse by imperial Christianity, Rome may have ruled but as the Roman poet Horace (65 BCE–8 CE) observed, though the Romans conquered militarily, the Greeks conquered culturally. And so the Museum continued to flourish until the Christian Emperor Theodosius proclaimed in 369 CE that Alexandria's pagan temples be destroyed, and a mob set it on fire.

Over these roughly six centuries in which the Museum flourished, inheritors of the Greek cultural tradition document a significant change in their experience of light. Theirs was fundamentally an active light, whereas the Greeks had regarded light as largely static. In the subsequent age of alternating calm and storm, of tremendous flux as Gnostics and Christians, followers of Mithra and Mani, polytheists and atheists competed for converts, a light defined by action was perhaps to be expected. We saw in the last chapter Athenian Luminaries explore the complex but rather fixed psychology of light; in this chapter, we'll see their Roman heirs explore its origins and active development, giving it a biography. They witness new acts of light, which prompt an investigation of its history, and find a language to express their experience by interpreting the writings of the sages once their classic texts were established and invested with authority, or "canonized." These included Homer and Aeschylus, Plato and Aristotle. And Roman scholars like the prominent senator Cicero (106–43 BCE) devoted themselves to their interpretation.[1]

The works of Luminaries like Plato or, for Jews and Christians, the writings of Genesis were refracted, then, through the lenses of Roman interpreters. By this method, writers who witnessed new acts of light found expression for them. Thus I call these men "Refractors"—Lucretius the philosopher-poet (99–55 BCE) yearning for Greek certainty and objectivity in his experience of an external light, and John the Evangelist (c. 90 CE) desiring an intimate relationship with the light he encounters, whereas Plotinus the Neoplatonic philosopher (204–270) and Augustine the Christian Bishop (354–430) plunge into their experiences of absolute light. But all are obsessed by acts of light, and all these Roman writers, with the exception of Lucretius, reinterpreted the works of Plato and Genesis, finding language for their encounter with light. Platonism, particularly its apprehension of light as a fundamental reality of the universe, mediator of the eternal and earthly realms, pervaded the mental atmosphere of the age much as Freudianism does ours. No one was immune, not even biographers of Jesus like John and Christians like Augustine.

At the same time, artists found a language to express their experience of light in mosaics and frescos. Reinterpreting their predecessors, such as Apelles, they introduce the age to an active, independent light in contrast to the Greeks for whom light is bound up with its contrary. Romans experience light not in relation to shadow but as a free agent, having an autonomy that ultimately arouses curiosity about its past, which prepared the way for writers who give us our first glimpse of its biography. As an agent, light has a history of actions in our world. And early in the Roman Age the writer Lucretius found materialist expression for this worldly light. These early Refractors anchor us in the sheer physicality of a self-confident Roman light. The later writers construct systems from their experience of light's activity in order to change souls, as did John in his Gospel, or the body politic, as did Hadrian in his Pantheon; or construct biographies of light to explain their profound longing (Plotinus) or guilty resisting (Augustine).

BRILLIANT MOSAIC SKIN

Ironically, Alexander the Great made Rome possible and in doing so preserved a unique document of light. The break-up of his empire, as his generals like Ptolemy I quarreled and divided up conquered territory after his death, provided an opening for Rome militarily. Though past its height, it continued to pursue conquest and domination within Lucretius' lifetime (first century BCE). One significant reward of victory was plunder. As it poured into Rome from all over the Greek lands, Romans were introduced to Greek art such as paintings from Macedonia. One such treasure was a depiction of Alexander at the climax of a decisive battle with Darius, King of Persia, possibly by Alexander's designated painter Apelles, whom we met earlier. Originally created about 330–320 BCE, this was brought to Italy in 146 BCE.[2]

The Alexander Mosaic Artist's Encounter with Light as Light

By a great stroke of luck, forty-six years later a mosaic inspired by this painting, which has not survived, was made in the House of the Faun at Pompeii (see Figure 5.1). The mosaic with its life-size figures showing Alexander (left) in battle against Darius (right) retains its energy and vividness to this day. Unlike paintings from earlier periods, mosaics preserve over the centuries the original vivid lights and colors of their small, roughly cut cubes, which were set in a bed of fine mortar and then grouted, ensuring their survival. Even today we are dazzled by their brilliant "skin" that spreads across the inside of the room. And so we're privileged to see first-hand an individual Roman artist taking pains to depict light itself as he encountered it, a light intruding into the scene in a way that we've not observed before in Greek or Egyptian painting. This mosaic goes beyond depiction of the highlights we know from reports of Apelles' paintings, which marked an important development in the human experience of light. We witness here a change in the artist's relation to light; we're privy to his experience of light as light, what the art historian David Summers calls the "optical light" of the Romans. We are made aware as viewers that we are viewers witnessing acts of light in the world.[3]

Thus the character of light in this mosaic appears to vie in importance with that of Alexander himself. While the artist portrays remarkable psychological complexity in his work, capturing the reactions of Alexander and Darius at a crucial turning point in the battle, he takes equally great pains to show the light refracted in the dust of battle in the background and the fall of light in the foreground. We're made aware, as we stand in front of the mosaic, of light entering the picture from over our left shoulder. Meticulously recorded by the shadow of the horses' left rear leg in the foreground just left of center, the artist exhibits a new self-consciousness about natural light and how it falls. So deliberately does he give us a depicted light, such as that falling on Alexander's head, that one viewer has deduced the exact time and direction of Alexander's charge.

Reinforcing this is the attention paid, rather incongruously in the middle of a momentous and gripping scene, to reflection. Astonishingly, the fore-

Figure 5.1. Battle between Alexander the Great and King Dareios, c. 100 BCE
Museo Archeologico Nazionale. Erich Lessing / Art Resource, NY

ground is dominated by a fallen soldier's shield, which we can't miss because its form is echoed by the chariot wheel behind. The artist delights in reflective surfaces as they reveal the nature of light as light to him, so front and center he presents us with a most magnificent reflection in the polished shield under Darius' chariot wheel; in its convex mirror the bright face of a fallen soldier.

Clearly, by emphasizing reflection, insisting that it occupy our view, the mosaic artist calls our attention to something new in his experience: that of light declaring itself. Taking great pains to show the white highlights of the sun's reflection from the mirror's rim and, more significantly, the soldier's reflection off its convex surface, he foregrounds light as light. Going beyond Apelles' use of highlights, those we see on the back of the black horse and on the rim of the mirror shield are not incidental but support the drama of reflection, compelling an almost hyper-awareness of light.[4]

Equally important, as a window to the artist's unique experience of light, mosaics of this period (first two centuries BCE) were not generally public works but installed in the floors of private houses belonging to the wealthy, as was the Alexander mosaic, and then mainly in dining rooms so guests could enjoy them. This meant that they reflected the personal experience of their patrons who with their hired painters introduced new genres, such as the still-life (*xenia*), the landscape, and the "unswept floor," which humorously depicted table discards like nut shells and fish heads. Mosaics became so popular that they commanded higher prices than the largest works of many painters of the time.[5]

Dioscurides' Light in Action

The senator and renowned writer Cicero hired the artist Dioscurides of Samos to create a mosaic in his villa in Pompeii based on a scene from a popular comedy of strolling musicians, which had been represented in a well-known painting some years earlier (see Figure 5.2). In this radiant mosaic Dioscurides reveals what is essential to his experience of light (and perhaps Cicero's). As we would expect, he masterfully deploys modeling (*skiagraphia*) and perspective (*skenographia*), which Apollodorus the shadow-painter and Agatharcus the scene-painter, respectively, had introduced centuries earlier. We see this in shadows cast by the players on the floor and the wall, as the sun illuminates them from the right of the scene.

And going beyond this expression of naturalistic light, Dioscurides employs Apelles' invention of highlight as well, what Pliny called "splendor" (or "gleam"), to achieve an expression of optical light. We see this in the gleam off the curve of the tambourine and the luminous reflection of sunlight off the wall behind the musicians, making it glow as if excited by the music. But the artist goes further, revealing the heart of his newly experienced light that finds expression in the depicted light of this magnificent mosaic.[6]

Dioscurides shows us light in action. His sunlight alters colors dramatically, such as the bluish-green in the tunics of the men and the flute player turning golden-yellow under the direct light. And this works in conjunction with the large cast shadows on the ground and wall that place each musician in the setting.[7] Each possesses its own space in clear relation to the larger outdoor scene. Revealed in this exquisite mosaic, then, is an active light exploring space and revealing forms and, more important, in its depiction making us aware of the artist's experience of it as active. It does things in the world. That is, Dioscurides expresses his experience of an active, instrumental light. He captures more than aspects of its character, which we've examined thus far, but its actions in the world. This is new.

The Greeks like Thales and Parmenides experienced an aspect of light's character that enabled them to employ it as an instrument, but they did not experience light as itself instrumental, acting on its own apart from them. This shift from exploiting an aspect of light's character to experiencing the actions of light parallels a shift in the understanding of seeing itself. Plato, like Homer before him, understood this act as fundamentally centered in our

Figure 5.2. Dioscurides of Samos (1st BCE). Musicians. Street scene. Mosaic from Cicero's Villa, Pompeii. 1st century CE. *Museo Archeologico Nazionale. Scala / Art Resource, NY*

eye, which moves out into the world. But after Plato the act of seeing is centered in the world, which enters our eye as images.[8] With this displacement we see a shift from the Luminaries' testimonies of light's character to the Refractor's revelations of its action. For Plato the sun as an image of the Good was static, as we have seen; for the Refractor of Plato like Plotinus the sun's rays as an image of the Mind-proceeding-from-the-One are active. His concern is not character but the acts of light. And ultimately entering the body, as we'll see, these become an existential experience.

This active, instrumental light is common to the Refractors' experience from the Late Republic (c. 145–44 BCE), as Rome aggressively and brilliantly exploited the Mediterranean world, to the dissolution of the Empire in the fifth century CE. We'll witness in this chapter specific acts of light at the core of each of the Refractors' experiences. The "glorious" and "mysterious" light that Reverend Hunter savored is all the more glorious for its active power in the world around us, all the more mysterious for its power within. They are acutely aware of this instrumental, optical light in the world and express it in core images, such as the jagged flash of lightning in a rain-dark sky (Lucretius) and the sun's "pure shining forth" after a storm (Philo) that illuminates the very intellect. It penetrates their being with enough force in some instances, such as Plotinus and Augustine, to change lives.

These core images, then, are not things lodged in their experience like a gemstone in its matrix but acts of consciousness that meet acts of light. A pure product of their apprehension of sensual reality, the image results from the fusion of inner and outer experience in their imagination, which cannot be separated. The image in turn exerts a pressure that finds expression in novel bits of language, metaphors used without explanation and requiring none. They do not mean anything other than how they are used in a specific, individual testimony.[9] Unlike metaphors, symbols rely on conventional language as they attempt explanation and are always double, being both themselves and signs pointing beyond themselves, as Plato's sun was both our familiar sun and a symbol of the Good. In moving from image to metaphor, a witness to light like Plotinus moves from experience embodied in his core image, "light is transparent to light," to language embodying his experience. But all testimony to the acts of light starts with the image. It is the lightning rod that conducts an encounter with light into the body.

Over time metaphors and symbols, which gave expression to specific acts of light, often become so much a part of the cultural conscience that they harden into clichés, as did Plato's sun-dazzle into the Absolute. At the same time, metaphor can break open language and reality once again for a new experience to emerge.[10] Aware of metaphor's power, some witnesses like John the Evangelist wrestle with the literal-metaphoric force of light in their lives, its psychic activity he calls "logos," whereas Plotinus celebrates its ability to unify the literal and metaphoric. Others start from their "extraordinary longing for the rays" of the sun to "penetrate" them, shattering all metaphoric scaffolding, or after experiencing the "light of full certainty" end, as we'll see, by denying any distinction between the metaphorical and the material.

Villas like Cicero's give us images of naked, optical light in action. Complementing their mosaics, these houses also host a flowering of fresco wall paintings from the first century BCE through the first two centuries CE that speak to us with extraordinary directness, bearing witness by their metaphoric force to an instrumental light. Their images strike deep into us. And so they should as individual expression in contrast to the public art in Rome at the time, such as sculptural portrait heads of statesmen and monumental architecture celebrating the power of the Empire, which given their propaganda purpose, reveal little personal experience of light. For this we turn to the private villas and their wall paintings with their landscape and still-life images.

THE LIGHT OF THE WALLS

It was Spurius Tadius "who first instituted that delightful technique of painting walls with representations of villas, porticoes and landscape gardens, woods," reports Pliny the Elder in his *Natural History* (77–79 CE), in fact, "every sort of thing which one might want." The villa wall paintings found in Rome and just outside on the Bay of Naples in Pompeii and Herculaneum confirm this claim about their delight and range of subjects. These frescoes shone with a unique brilliance, a luminousness that the artist developed a special technique to achieve. In some villas murals covered the entire wall of a room, their images unwinding like a banner. Pliny continues with a bit of sly humor, Tadius "also began the practice of painting representations of seaside towns on [the walls of] open galleries, thus producing a charming view with minimal expense."[11] Poor villa location? Hire an artist.

Interest in landscape was a Roman innovation that coincided with their experience of an active light. And this marked a turning away from the classical preference for architectural settings. As landscape became a subject of attention in the reign of Augustus (63 BCE–14 CE), so the light of nature does as well. We see in these paintings a new attention to light and its effects as it reveals the contours of hills or the outlines of porticoes by the sharp contrast between surfaces in direct light and those in shadow. This new subject of landscape—natural light—also shows up in the still-life paintings that appear on the walls, such as the intricate workings of light revealed in a bird's feathers. Thus wall paintings exhibit not only the optical light in action we glimpsed in mosaics but a light that comes to regard its own acts in a sense. Painting by its very nature "engages in self-reflection," shining a light on its own act of light, its depiction of light, inviting individual and social reflection.[12] Consequently, with the rise of painting in this period Roman artists become self-conscious in the presence of light's behavior. And projecting their self-consciousness, this new a new kind of light in the lives of Romans becomes light's "self-awareness."

But we can't understand these paintings and this light apart from their settings. Fresco images filled the Roman houses established in republican times during the first centuries BCE and CE. These often participated in an ensemble of wall paintings coordinated with the villa's architecture and its orientation to the sun.[13] Entering a typical upper-class urban villa in the second century BCE, such as the "House of the Tragic Poet," guests find themselves in a narrow hall or vestibule, which could rise upward over twenty feet. This leads to a formal entrance hall or atrium that opens into a large room, its floor dominated by a glittering mosaic, and beyond onto a colonnaded garden (peristyle), the largest and most important space in a Roman house. Along the wall of the room facing the peristyle a painted garden scene on occasion seems to blend with the owner's actual garden.

Clearly, this house is more than a dwelling; it is a display of status, both a projection of private needs and a product of Roman society. And central to this display are its wall paintings, which the Roman architect Vitruvius, writing about 15 BCE, claims were integral to the house. They are extensions of the villa's owner who commissioned the works and with a hired artist determined their composition and position within the house. The owner worked with the artist not only to determine the patterns of images, including their subjects, but to coordinate them within the space of the rooms, striving to achieve an ordered relationship of architecture and painting. Owners, however, some of whom were even artists themselves, made the final decision. These paintings are to a large degree, then, personal expressions of the patron and artist, the fresco's depiction of light being their experience of it.[14]

A Shifty Character and the Dissolving Act of Light

Typical of the more personal art of Roman wall paintings were those in the house of Publius Fannius Synistor, a sumptuous villa located in the Italian town of Boscoreale just north of Pompeii in a farming area called the Campania. Built just after the middle of the first century BCE, the villa was buried by an eruption of Mt. Vesuvius on August 24, 79 CE, a disaster for him that happily for us preserved many of his frescoes, which featured landscape and still-life images. Though they borrow from Greek art, elements are combined and used in a distinctively Roman way. The wall paintings of Synistor's villa, like much Campanian fresco, reveal a "conspicuous" or depicted light that brings out surprisingly illusive and mobile effects. On his bedroom wall, a cluster of buildings is washed by a strong Italian light. Here guests experience natural light. Intense sunlight plays on the houses among scattered patches of shadow. Then struck by gleams and flashes the landscape ceases to be solid. It ceases to be passive and becomes by fleeting acts of light, kinetic and unpredictable.[15]

These effects are alien to Greek art, which was devoted to a rational objectivity. The flaring lights we see in Roman wall painting are intended not to reveal any objective reality but discrete appearances, images of light acting in specific moments and spaces. A wall painting that depicts villas built in the next century takes this capturing of light's momentary actions further. These "Houses at Noon" stand in a "quivering haze of broken gleams," as one viewer describes it, that feels like "the fabric of a dream."[16] This light is a shifty character. It reveals an energy roiling in the light on the buildings in Synistor's bedroom, where an active light cooperates with us as viewers. The painter expresses a fundamentally ambivalent experience of light embodied by a mobile landscape that we normally assume to be static. Light here acts to disorient us, as it does visitors entering the House of the Faun, host to the Alexander mosaic. The walls seem to drift in and out by this depicted light, as they come face to face with an "outside" that is "inside" and at the same time an "interior," the atrium, that is "outside." Some spaces open to the sky some enclosed, its wall paintings with their outdoor garden scenes in natural light smear the boundaries of a peristyle's actual garden open to the sun.

Fannius Synistor's guests would have been particularly alive not only to the setting of his villa, the sunlit view from where it is situated, but to the flow of light and space from room to room. This flow with its changing interior and exterior views, as it constructs the visitors' perceived space-light of the villa, would necessarily be as ambiguous as the light of the paintings that guided them. This depicted light of fresco reveals an aspect of its character foreign to the Greeks who experienced natural light as definitively giving them objects outside themselves in the landscape. The Roman painters experience a different light. They are nagged, it seems, by a sense of having grasped objects before being fully aware that they exist outside them. It's not emotionally evident whether objects take on their reality in the external landscape or in the viewers' internally apprehended world. In short, the light acts in both arenas. This acts to break down the threshold between what the Greeks assumed to be objectively external and internal. And such inherently ambiguous action renders Synistor's experience of light's character ambivalent.[17]

The ambiguity of the light in Synistor's wall paintings reinforces the handling of spaces within a house designed, it appears, to dissolve boundaries and deny the limits of actual spaces, the floor, ceiling, and walls of their small rooms. All "real" spaces become the "fabric" of a "light-space" dream. As we enter one of Synistor's bedrooms from the peristyle, which we can do in the reconstructed room of the Metropolitan Museum of Art (New York), we see on the rear wall a "spectacular assault" upon these spaces in the painting *Garden with a Grotto and a Pergola; Landscape on a Yellow Ground*. Here the simulated space of the wall painting's cave and pergola vies with the real space of the window with which the painting was integrated. And working with paintings on adjacent walls, the simulated takes over from the actual space of the room in a most compelling way, as what is painted draws on components of the villa's site itself. This preoccupation with thresholds between the real and the fictional, which the new light highlights by erasing, also satisfies a personal need of Fannius Synistor. He is typical of Roman patricians on their country estate escaping the boredom, as Lucretius wryly notes, that propelled them back and forth between the city and the country. Being mobile, the light of Synistor's experience and that of his friends transforms actual boundaries into the fictional and limitless "fabric" of a dream.[18]

The wall of an underground dining room in the Villa of one of Synistor's patrician contemporaries, Livia Drusilla, wife and advisor of Augustus, built in Prima Porta just outside Rome (c. 20 BCE), ushers us deeper into this dream. Here a lush painted garden seems to overwhelm and incorporate any actual planted garden. Its simulated light appears to drift down from the real garden in the peristyle above or will shortly rise up into it. A misted, gentle light pervading this garden, which does not shine on any single object but lives with them as an equal presence, a light in other words, participating in the actual and the fictional. Visitors find themselves inside a grotto, a natural cave, as the rough outcrop at the top of the painted wall indicates, looking out at a walled garden. By placing the painting underground and the garden at a distance from the observer, however, the artist convinces them that they are outside. As a result, they are directly involved in the ambiguity of reality and fiction here, just as was suggested by our viewing of the grotto on the rear wall in Synistor's bedroom. The garden mural "dematerializes" the surrounding walls, opening the room in every direction. The light that unites fantasy and reality here is complicit in our disorientation. And these painted images document the light experienced by Roman patricians, such as Livia and Synistor, and their artists as it acts to dissolve the very boundaries of their villa walls.[19]

Still-life and the Subversive Act of Light

Not only in a garden scene is this true, and there is nothing like this in Greek art, but as it acts in a still-life as well. On the wall between the window and the grotto scene in Synistor's bedroom we see the image of a transparent glass bowl with fruit. The painter self-consciously paints this touchable glass bowl as inseparable from his depicting light itself. The bowl's reflecting and absorbing the light makes this clear. Transparent glass, a recent invention as Seneca reports in 65 CE, was in itself the ideal exhibit for Synistor's experience of the new mobile light acting to erase thresholds. Roman artists were drawn to this invention, an inherently ambiguous material, solid yet transparent, miraculously responding to light.[20] And they take sheer delight in this miracle play of light, as we see in a still-life painting of peaches and water jar in a villa at Herculaneum (c. 50 CE) not too far from Synistor's. We're struck by the planes of light that form the shelves, which support the peaches, as well as by the brightening ends of the shelves, which the graceful curve of the peach stem frames (see Figure 5.3). But the water jar, where light pools and plays, captivates us totally, as it did the artist who brings it forward by providing a shadow backdrop. He makes light fairly dance on the surface of the exquisitely transparent glass and the water as if they were partners.

And in fact, they are partners in expressing light's mobility and ambiguity. Still-life paintings such as those on the walls of Synistor's house and in the villas of his friends crystallize the new experience of light. We've seen landscapes, including buildings and gardens, where this ever-on-the-move light dissolves actual boundaries between inside and outside spaces. Light collapses our sense of what happens within and without, our apprehending objects internally by imagination and understanding them externally by vision. The object seems to be at once inside and outside us. But still-life wall paintings intensify this experience by undermining the very act and technique

Figure 5.3. Still Life with Peaches and Glass Vase. Roman fresco from Herculaneum. ***Museo Archeologico Nazionale. Scala / Art Resource, NY***

of depicting light.[21] The artist as a maker of the figurative, painted landscape or still-life must deny it in representing light because in his painting actual light (the external light bounced from pigments) and depicted light (the internal light constructed by the mind) become one. When optical light itself becomes the subject, we become involved as viewers in a paradox: we are observing "real" light and "fictive" light at the same time. And we can't separate them in our experience. This drama is in part what excited Fannius Synistor and his contemporary patricians, expressing their encounter with light.

And it is this light in action that is characteristically Roman. As it emerged in the experience of a patrician like Synistor, nature became a subject of art and with it landscape where natural light insisted on its presence, no longer content to fade into the background. So light gets its own portrait. For the first time, light comes into its own as a depicted light, shifty and subversive, always moving and always erasing boundaries. Each of the Refractors, whose testimony to its acts we'll be hearing, is intimately familiar with this active and elusive light. Although Lucretius stands alone in his naturalistic assumptions, all share the experience of light acting in the world, either within them or within nature. Not bound to its source but autonomous, it becomes in their dealings with it a "self-conscious" light. Although "invisible," by definition a clarity that reveals objects, such as the still-life water jar, it nonetheless calls attention to itself as light. In the artist's experience, it is precisely this paradox, this strangeness in revealing objects while hiding itself, that brings it into their consciousness.[22] A strangeness we'll see that haunts later witnesses of light's character like Plotinus.

LUCRETIUS' SWERVES OF LIGHT

Just a couple of decades before the villa was finished (c. 40–30 BCE) that became Publius Fannius Synistor's in the first part of the next century, Titus Lucretius Carus (99–55 BCE), known simply as Lucretius, poet and philosopher, finished his famous epic poem "On the Nature of Things (De Rerum Natura)." In this he set out to free all men from their paralyzing fear of the gods and of death. Living in a time of civil war triggered by the renegade army commander Sulla, who had executed more than four thousand leading citizens, Lucretius admits that it is difficult to "sit calmly writing." Nonetheless, in his "country's hour of trial," as he says, Lucretius writes a poem of "tough-minded grandeur," one that reveals a fascination with natural phenomena, such as the sun "sowing the land with light" (II. 211).[23] And in the midst of chaos Lucretius trusts to the senses and reason, being skeptical even of the language of his own poem. His is an intellectual probing without illusions to explain the world in which he finds himself.

At the core of Lucretius' explanation lodges a single image borne, I believe, from two pivotal experiences of light, one outdoors and one indoors, that appear to have merged in his psyche. Lucretius spends a lot of time outdoors in storm and rain observing natural phenomena where he had an encounter with a light of stunning character. "When storm clouds gather," he says, "we see the lightning scatter" (I.761–62), and he observes that the lightning bolt cuts obliquely across the parallel lines of rainfall. Light reveals itself in an angled flash. A kinetic light, it is also an instrumental light, having brought fire "to earth to mortal men" (V.1092; thus no need for Prometheus).[24]

This experience is complemented by a second one that he has in the comfort of his study. He urges his friend Memmius, governor of Bithynia in 57 BCE to whom Lucretius' poem is addressed, to "watch whenever the bright rays of the sun / pour shafts of light into a darkened house: / you'll see a thousand motes [flecks of dust] a thousand ways / commingling in those very shafts of light, / engaged in battle and blow as if in strife / eternal, host against host, without a pause." He goes on to comment, and this brings his outdoor and indoor experiences together, "from this you may conjecture of what sort / is the endless movement of atoms in the void" (II. 114–22). Here Lucretius moves boldly from this natural phenomenon to an underlying law, pointing out the importance of Memmius's observing these "motes" that are "seen swirling about in rays of sunlight / because their swirl reveals that, all unseen, / under the surface, atoms are on the move" (II. 125-28). By bringing together the jagged way in which light moves as lightning and the helter-skelter way in which dust motes move in the light, Lucretius creates an image that drives his argument and lays the foundation of science.

His image brings together sensual apprehension and imagination. This becomes clear when we look briefly at how Lucretius' poem refracts the writings of his "master," the Luminary Epicurus (341–270 BCE), who posited atoms and the void and a boundless universe as the fundamental reality. "Basic particles," atoms, "survive unchangeable" (II.750–51) in the void, moving without purpose, randomly in a "limitless" space (II.92–93), which does not care about us. Yet if they "fall straight downward through the void / by their own weight," how do they interact to create the world our senses detect? Epicurus reasoned that at unpredictable times and points, they "swerve a bit" and change direction (II.217–20), resulting in interactions among atoms and their generation of combinations—the material world. Lucretius' own observation of lightning's random jagged motion and light rays' interaction with dust motes reinforces this teaching. He swerves from his "father" Epicurus, however, in taking the operation of chance within the void more seriously, as we might expect given the chaos around him. No god or mysterious power ordains the presence of light; an erratic character, it appears randomly in our lives.

This returns us to Lucretius' fundamental image of light, a kinetic instrumental phenomenon. His artist contemporaries, the wall painters in his friends' villas, experienced an ambiguous mobile light, but Lucretius experiences a radically active light inseparable from chance, one that is unpredictable and thus wholly impersonal. He moves out into the chaotic and turbulent weather of a storm to observe, and the lightning he sees jagging across the rainfall becomes for Lucretius, turning to metaphor, the "visible model" of the world as it is "realized in nature."[25] The motion of this particular light manifests the underlying reality of atomic motion. Bound up together, they are instrumentally linked. In Lucretius' imagination the kinetic movements of light and of the atom are intimately connected in the creation of our world, which he calls the "shores" or "coasts of light" (II.577, 617; V.782, 1455). Here "lightning . . . goads lightning on" (IV.190), and "death mingles with the wails" that babies raise at birth on "viewing the coasts of light" (II.576–77) under that "well-spring of bright light, the heavenly sun." This is an active sun that "endlessly floods the sky with newborn rays / as it supplants light speedily with new light" (V.280–83). Light's indeterminate swerve is the atom's swerve that generates the universe.

In this primary image Lucretius embraces an uncompromising "naturalistic conception of things," going beyond Akhenaten, a "great work of imagination,—greater," as the philosopher George Santayana reminds us, than any "dramatic or moral mythology." Lucretius' refraction of Epicurus' idea of the swerve appears to be driven by his imaginative fusion of light-in-action (lightning zigzagging) and action-in-light (dust motes moving helter-skelter in a shaft of light) into a single image. And this unflinching insistence on indeterminate motion, the chance interaction of particles, which rests at the beginning of science, brings the world into being. In his conviction that nature constitutes an impersonal creative-destructive spiraling, which the structure of his poem reflects, Lucretius exists in a world of light in action, continual flow.[26]

But acts of light for Lucretius occur at the atomic level, not the personal. At the same time, he forces us into the role of observers as well, precisely where the world of atoms in the void and their chance swerves leaves us. We can only observe the effects of atoms by the behavior of light. Light observed from a distance behaving erratically is thus not the predictable though ambiguous domestic light of his artist contemporaries. His is a wild outdoor light that can't be predicted. It acts on behalf of the cosmos not his world, its instrumental function in our world being literally an "accident," a swerve having nothing to do with us. He embodies this light whose actions are more unsettling than the wall painters' light in a poem that takes as its model the whirling storms and jagged flashes of lightning he sees. Like the character of this light he encountered, his poem, by which he found expression for his experience of light, is a "vortex" as the historian of science Michel Serres observes, that "turns endlessly back upon itself."[27]

Lucretius rejected the notion of any transcendent realm, where the later Refractors of this chapter—Christians, Gnostics, and Neoplatonists—assume it. In the Gospel of John, God (the logos) creates the universe; in the gospel of Lucretius, chance (the swerve) creates the universe. A truth revealed to him by his experience of a zigzag, helter-skelter light. But Lucretius' impersonal relationship with light, furthering Thales' tradition of understanding the universe, gives way to the personal relationships of Philo, John the Evangelist, Plotinus, and Augustine, all influenced by Plato in the tradition of aligning rightly with the universe. And in doing so they not only revalue the character of light, internalizing it, but for the first time in history explore its biography. Defined by its remoteness, for Lucretius this character could no more possess a history than the eternally existing infinite cosmos. But for Plotinus and Augustine a few decades later, light's personal overture fired a need to probe its biography, the history of its acts.

SYSTEMS OF LIGHT—JOHN'S GOSPEL AND HADRIAN'S PANTHEON

We saw Lucretius' movement from an image, his initial encounter with an erratic light, to his metaphoric interpretation of this core experience, his "model" of the universe, a familiar progression in our experience and imagination of light. Turning now to John the Evangelist and Hadrian the Emperor, we'll see them extending this by symbol and beyond, creating systems of interpretive relationships. In their effort to communicate the meaning of their light-experience, these can in contrast to metaphor be nearly limitless. Living a century after Lucretius, they respond to an autonomous light, neither the interactive light of the painters nor the distanced light of Lucretius. They wrestle with the literal/figurative "and/or" character of light, coming to terms with it in systems of language and architecture, respectively, until they are able to conscript it in a mission inspired by light's character in the first place. John partners with light to heal the world and bolster his authority, Hadrian to heal the nation and consolidate his power.

John's Gospel

John the Evangelist, "writer" of the Gospel, opens his book with a hymn, refracting Genesis, "In the beginning was the Word [*logos*], and the Word was with God and the Word was God" by whom all things were made by. And "in him was life; and the life was the light [*phos*] of men. And the light shines in darkness; and the darkness comprehended it not" (1:1–4). Part of the existential ground base of his book, this hymn sounds notes that reverberate throughout, namely, John's experience of a light uniquely powerful, beyond the natural light of a star penetrating night or the sun dispelling its darkness. Furthermore, John apprehends this light as a person, the Logos who reveals himself to himself by self-questioning. "Logos" has nothing to do with metaphysics and everything to do with physical being. That is, the hero of John's book questions himself and his actions. And John's encounter with this and/or character proved pivotal. For he viewed the living itinerant preacher who shared meals with all and dispensed free healing—Jesus—as the incarnation of light.[28]

Having encountered the incandescent Jesus, light becomes in his imagination instrumental, a being who created the world and now acts to save John and the world if his claims are believed, giving it immortality. We're tempted to dismiss John's statement about this character, "In him was life; and the life was the light of men," as a metaphor, a mere figure of speech that he uses to express its saving action in us. But John insists on this light (*phos*) being literal. When he states that "God is light, and in him is no darkness at all" (I John 1:5), he means this literally. A truth difficult for us to comprehend two thousand years later, but one Augustine who followed John by just a couple of centuries understood, commenting, "One does not say that Christ is light in the same sense that he is called a rock, for the first statement is literal, while the second is only a figure of speech." It is the immediacy of solar light in John's experience, I suspect, that demands his using "light" literally but rock figuratively. Because for John the Logos is not just a mediator between spiritual and material worlds but their identity, John wrestles throughout his book with metaphor, vacillating between a literal and figurative light. The arena of light in which John operates, however, ultimately demands that the literal triumph. Thus John uses the nouns in his hymn, "Word" (logos), "life," and "light" literally.[29]

He grounds this triumph by giving light a biography. Light is inseparable from the history of logos, the story of creation. He must go back to the beginning. Thus he refracts the opening of Genesis, interpreting and rewriting it to accord with his own experience of light. Before Logos, light existed; "And God said, let there be light: and there was light" (Gen 1:3). The word existed as light before being sound. So light is born, God the absolute light who is at enmity with darkness; grows into Life (1:4), the Logos, and ultimately becomes flesh and lives among us (1:14), the Light/Word/Jesus. These three stages in the life of light are a literal history. And primordial light in the beginning is thus the present light in experience.[30] For John this biography of light is central.

But John in the first century CE must find the means of expressing his radically existential experience of light acting in his life and in the world. Accordingly, he invents a system of rhetoric that enables him to come to terms with the light he has fallen in love with incarnate in the incandescent Jesus. His system, an elaboration of his core image of the light-being having a personal history, provides a framework for understanding such a unique figure.[31] It is a system, evidently, derived from his experience of natural light that gives us the multitude of differences we depend on, such as between trees and rocks, but also erases these differences. John's vacillation between literal and figurative reflects this basic observation. Objects that appear separate merge in changing light and shadow. But we find this merging to be figurative when we stumble over one. Fundamental for John is light's ability to reveal differences in the world while erasing them.

Extending this perception, he constructs a "light system" of language that the New Testament scholar Norman Petersen describes as one of "differentiation" and "nondifferentiation." Light's behavior in our world flows from its uniting and dividing nature, embodied in Jesus' basic mission of salvation. He separates darkness from light in order ultimately to erase this division by the action of Light itself, which is inseparable from its biography. For in the beginning, as John rewrites Genesis, light is undifferentiated. In his opening hymn Logos and God, life and light are an identity. But when the Logos enters the world, which "was made by him, and the world knew him not" (1:10), light is differentiated, "the darkness comprehended it not," only to reclaim identity again in the salvation of the world from darkness. This tension, reflected in the wrestling of literal and metaphoric language throughout, reveals the complex character of Light/Logos. Thus his Gospel's light system, a flexible and powerful rhetoric enables John to come to grips with the personage of light, the "luminosity of the Word/Light" as it appears in Jesus (1:14) and acts within him.[32]

Hadrian's Pantheon

While John was writing, an old and unremarkable temple in Rome burned down (80 CE). At the time no one could have imagined what was to replace it some three decades later when the Emperor Hadrian (76–138 CE) and his architect designed and constructed the radically innovative, glorious Pantheon. Born four years before the fire, Hadrian became emperor in 117 CE, and having a passion for architecture, he recruited one of the greatest unknown architects in history to collaborate with him on the Pantheon (c. 117–127 CE). Though Hadrian undoubtedly drove his project's conception and creation, together they created a building dedicated in 127 CE that remains one of the most important in the history of world architecture. It is a monument to the unifying power of light. And by a great stroke of luck it has been preserved virtually as it appeared after it was first built.

From the outside the Pantheon appears rather nondescript, simply a marriage of the traditional temple front and domed rotunda. But entering it reveals another world. As the bright open court gives way to a somewhat dim porch, visitors' eyes adjust. Then passing into the massive domed rotunda, a blaze of light greets them. Standing on the floor of the dome in a magic circle whose diameter exactly matches the distance to its ceiling, they are transfixed by a shaft of light from a round opening in the ceiling (oculus) piercing the cavernous space—a space that is to this day as a modern architect testifies, "totally suffused and inhabited by light."[33]

In this building the Romans turned Greek space outside in. The Parthenon, built over five centuries earlier in Athens, reveled in its exterior, varying its marble fabric of brightness in the open air. The Pantheon by contrast delights in its interior, manipulating the light that pours through its oculus, which Hadrian knew he could use like a lens to vary the quality and movement of pure natural daylight. In doing so he presents light as light, a palpable body of light living in its own house. We are reminded of Newgrange and its shaft of light, a singular semi-annual occurrence presided over by the seer. But Hadrian's, by contrast, is a regularly available active light, free of a mediator and open to all. In short, the Pantheon was "made for light." Caught up in our exhilaration at its sheer beauty, as we ascend in spirit through the oculus, we are hypnotized in place by light that invites us to touch it even as it eludes us.[34]

At the same time, the play of interior light includes shadow that energetically questions the light shaft. As it angles in from the oculus, light touches the nested rectangles of the wall coffers, which display intricate patterns formed by subtly shifting gradations of light and shadow breaking towards and away from us. And we find ourselves in a theatre of light. The dramatic tension of light and shadow brought into a unity stirs the emotions in ways that are powerful and elusive at the same time. A tracking spotlight, in effect, focused by the oculus heightens this effect. Its large disk of light moves slowly across the interior wall of the dome as it marks the earth's turning. Tracing a parabolic path through the northern half of the building, it moves throughout the day, starting high on the wall in the morning, dipping to the floor midday, and ending high again late in the day. We cannot take our eyes off this disk. We delight in the drama around us taking place in the atmosphere of light that pervades the dome, but this continually moving and shape-shifting circle of light is the star attraction. Hadrian subordinates all the elements of the Pantheon to this revelation of the sun, an astonishing heliophany. Light, then, was the "prize for which all this concrete was poured."[35]

Hadrian, in designing and building the Pantheon, consciously constructs a light system. He shapes his building with the express purpose of manipulating a "great shaft of illumination," this always visible, "almost tangible rod of solar effusion," in the architectural historian William MacDonald's phrase. The viewer is alternately pierced and transported by its rod of light.

And Hadrian shapes it with the purpose of taking us beyond its concrete and brick. The total working together of the light streaming from the oculus and the play of light and shadow in the niches of the coffers as well as wash of light across the pavement below dissolves the building's materiality. This results from the building's "perfect unity of the light" that impressed one nineteenth-century traveler. We forget that the massive walls surrounding us and foundation beneath are mountains of physical material. These effects of Hadrian's heliophany, transcendence, and immateriality, which evoke in us an undefined but powerful sense of unity, manifest his expressive purposes.[36]

Creating a powerful heliophany, Hadrian celebrates a singular act of light. It convinces worshippers of their participation in a larger unity. As an emperor who has battled for territory and ceded territory in the interest of the empire, and witnessed religions in other lands fade and at home brighten, he no doubt realizes that he is on the threshold of history. The classical world he yearns for is gone, a new world arriving. For Hadrian, the greatest threat to the empire is dissolution, and he dedicates himself to preserving it. To this end he harnesses the heliophany, his experience of light acting to unify. The oculus sun stands fixed at the center of the world; its shaft's spotlight tracks the cosmos that revolves around the great dome. And the Pantheon's very structure is an architectural metaphor for the ecumenical unity of the empire. It is, further, a system of concrete and light intended to "unify unities," bringing into reality the universal. Nothing short of unifying the gods and state, the people and the state, to ensure the eternal existence and functioning of the empire is Hadrian's purpose. The Pantheon, the world's temple, metaphor extended to symbolic system, captures the act of light in creating "order, peace, harmony, unity."[37]

Having seen the light of Britain and Turkey, of Spain and Egypt, the unifying power of light, I believe, profoundly affected Hadrian. For after his extensive travels Hadrian identified himself with Phoebus, the sun-god who reigned over all lands. This god embodied the projection of Hadrian's fundamental image of light as unifier. And he gave this emotion physical form, projecting it into a statue. When he and his architect entered the rotunda on dedication day, they stood before Zeus-Jupiter-Helios, the supreme god, ally of the Great Sun, which had been placed inside. According to the ancients, as Hadrian knew, the sun is the "eye of Zeus," and in his Pantheon the oculus light is the epiphany of this greatest of the gods. The merged oculus light and sun-god held a supreme position because of light's uniting power. Just as Akhenaten's response to his experience of the diversity of human communities was the Aten, Hadrian's analogous response was the Pantheon.[38]

Worlds apart in status and power, John brings the Logos/Light into history for the world's salvation, Jesus intersecting time; Hadrian takes the Helios/Light out of history for his empire's sake, transcending time and space. Thus unlike the evangelist's system constructed from the rhetoric of language to express the paradox of Jesus-Light intersecting his life, the emperor employed the rhetoric of forms not to differentiate those in the light from those in the darkness, but to unify, reflecting a very different goal. In doing so he furthers the exploration of Synistor and his painter who succeeded a century and a half before in creating monumental murals depicting light that attest their experience of a light as light and in which, paradoxically, real light and fictive light merge. For the Emperor Hadrian and his architect go a step farther and give us in their marvelous building simply light as light that dematerializes the building even as the light itself materializes in a heliophany.

POIMANDRES SAYS, "THINK ABOUT THE LIGHT AND UNDERSTAND IT"

Dedicated to all the gods, Hadrian's pantheon (*pan* + *theon*) expressed a desire for inclusiveness that extended to the God of the Jews and the gods of the Greeks and Romans as well as those of the rising Christian and widespread Gnostic sects, which were well established by Hadrian's time. John had written his Gospel, in fact, as an answer to a Gnostic-flavored work, *The Gospel of Thomas*. Both writers attempt to deal with the luminous person of Jesus but out of different experiences of light. Thomas, writing in the Gnostic tradition, says that light shines through everything, his Jesus declaring, "It is I who am the light" before the world was made, "Split a piece of wood, and I am there."[39] John's book counters the second assertion. For him logos/light is available solely as a personage, a light with a biography who was "made flesh." And the only way to appropriate this light is by "belief," establishing a personal relationship with the Light of the World not unlike a lover's.

Gnostic light, by contrast, is accessible through human knowledge, a self-discovery central to the Gnostic tradition. Flourishing before Jesus and popular everywhere in the Graeco-Roman world, each of the various sects of Gnosticism offered salvation through self-knowledge. What caused suffering was ignorance, not sin. This meant that all men could find the light within themselves, not merely enlightenment but the literal light itself that existed from primordial times. And they could look to a profound experience of this light for rebirth and freedom from suffering because of light's origins. Refracting the book of Genesis, as John also does, the entire Gnostic tradition is rooted in this history.[40]

This is a history that Thomas appropriates in his Gospel, for Jesus tells his followers when people ask where they come from, to answer, "We came from the light, the place where the light came into being on its own accord and established itself." In contrast to John's, this creation story places light within us from the beginning. We do not appropriate a light that walks beside us, but revive and liberate the light that resides within us. This intensely personal experience of light woven from knowledge of its biography is the essential Gnostic gospel. Our "luminous body" inherited from the primal god, which is the cosmos itself, acts powerfully within us for our own liberation from suffering.[41]

In the pagan Gnostic treatise *Poimandres*, Poimandres appears in a vision to a man named Hermes. In this vision, he experiences the power of his mind in its fullness as light and gives us an account of his liberation. It begins one night when Hermes "began to think about the things that are" and his thoughts soared "exceedingly high." But at the same time feeling his senses "held down by sleep like people weighed down by overeating and weariness," he sees a "vast and boundless" being approach him who calls his name and identifies himself as Poimandres. He says, "I know what you want and I am with you everywhere." Then suddenly in a "flash everything opened before me," Hermes testifies, and "I saw an unbounded vista. All was Light, a soothing and happy Light. And as I gazed I was entranced." But he is also liberated, saved from a "stark and terrifying Darkness" by the "Light a holy Word [*logos*]," which he hears as the "voice of light." Then Poimandres explains, "I am that Light . . . and I am the Mind." After Hermes thanks him for this revelation, Poimandres advises him, "But think about the Light, and understand it."[42]

Understanding light means knowing its history. This is why Hermes relates the basic facts of its biography: the "Mind" or fullness of light (pleroma) existed before all, then appeared Darkness and the fall into material creation, and finally emerged the saving "luminous Word" that Hermes hears within, verifying his vision. Earlier in the *Corpus Hermeticum*, which includes the *Poimandres*, Hermes tells us that this Mind the "Father of all, He who is Life and Light, gave birth to Man, a being like to Himself." That is, he interprets his encounter with light as not merely an act of self-therapy but an encounter with his true inheritance, a literal primordial experience of light. Thus by means of an ascent through levels of knowledge, anyone can reach back into his own actual history, which is light's history, to reclaim his genuine humanity. This means having gained true understanding—"all things are filled with light," including us.[43]

Hermes radically internalizes the light he experiences in the natural world. This is not a metaphorical "inner light," spiritualized in later ages. For it possesses a body. Because we know light's biography, we know that this is a literal light given to us at birth that in order to alleviate our suffering we need only access through self-knowledge. To understand the light, then, is to know our true self—the Light-Body-Mind. It is our essence. By his vision of Light who reveals its biography, Hermes can "enter into Life and Light," as he says at the end of the *Poimandres*, and break through his "weariness," his senses now alive and not "weighed down." He can see the true nature of light and its history, which is his own.

BIOGRAPHIES OF LIGHT—PLOTINUS' FLASH AND AUGUSTINE'S SEDUCER

An intense encounter with light, as we have seen, creates great pressure to make sense of it. We want to know what it means. Thus we turn to interpretation, discovering a way to deal with the significance of a specific experience of light. Drawing on their great powers of imagination, Refractors, from mosaic and wall artists to philosophical (Lucretius) and religious (John and Gnostics) poets, find language to express the meaning they have made. We've seen how John the Evangelist and Hadrian the Emperor in their search extend the core image at the heart of their light-experience via interpretative metaphor to imaginative symbol and beyond to symbolic system. The Gnostics interpret present light-experience as experience of light past, which they can recover by a right understanding of its biography. And turning to Plotinus and Augustine, we'll see how these writers extend their core image to symbolic biography, which grounds the profoundly personal meaning of their experience of light.

This creation of light's biography as the foundation for the significance of such an experience, however, begins some two hundred years before Plotinus with Philo (20 BCE–50 CE), a cosmopolitan Jewish theologian from Alexandria, Egypt, who refracted both Genesis and his hero Plato's works. Philo's core experience was intense: the "noon-day beam" of light "pours" into his "whole soul." And he radically personalizes this "torrent" of rays. They pervade his entire body, fusing the physical and mental. So powerful was this infilling of noon-day sun that in order to understand his resulting relationship with light, he had to ground it in the birth and history of light. Only by this story can Philo give expression to his overwhelming experience of light acting on and in him. He needs a larger context of significance in which to situate and make sense of what is essentially personal. For his is a profoundly subjective light, opposite of his older contemporary, Lucretius, who finds expression for the power of a stubbornly objective light. And given such a personal relationship with light, it is no wonder Philo pronounced light the "best of all existing things."[44]

Yet it results in "longing." The light he experiences at noon makes him yearn for the "unmingled and pure light" of God himself, he says, a light birthed before the sun. It is possible to apprehend this light because our mind is the very image of God, possessing his "likeness." This is why the internal world of our mind is, as he says and Plato agrees, "full of every description of light." And this mind of God within us shares the principal attributes of light. It is, for example, "invisible, though it sees everything itself." Just as "God, being his own light, is perceived by himself alone," so light is self-sufficient in knowing itself. Philo's is a self-conscious light in this sense, evading us even as it presents itself to us. It is this light acting in Philo's ordinary self that causes the longing within him and yet provides the power to satisfy this longing. It enables him to achieve the extraordinary self that enters the presence of God, the "pure" light.[45]

But he cannot do this alone. He requires a mediator between earthly "light" and the transcendent "pure light," emanation of the "Light-Stream from God." This is the "Logos," which John also borrows, that Philo by stages can appropriate to attain the "pure light" of God's presence. Convincing himself of the reality of this possibility, Philo engages in a bold refraction of Genesis. Rewriting the creation story, he traces the Light-Stream back to its origin and provides light with a history. And thus he grounds his progressive discipline. For him the biography of light is implicit in the creation story, saying that on the first day Absolute light emanates from God, and from this on the fourth day, material light emerges. This provides the basis for the stages of mingled light that he must traverse back to the pure "unmingled" light of God, progressing from natural light to mediating intelligible light (Logos) to the Absolute light. In effect, then, he retells the story of creation as the story of the birth and life of light, a history of the character light expressing itself in creation, which he recapitulates in his own life.[46]

Plotinus' Flash

No Refractors are more entranced by the sheer beauty of light than Philo and Plotinus. Trying to understand the light as Poimandres exhorts us, Plotinus (204–270 CE) enlists the help of philosophy, fusing Plato and Aristotle, but takes a remarkable turn. To help us understand how what Plato terms "incorporeal" or intellectual light can remain such, being pure "activity," as Aristotle maintained, even when it enters and reveals the corporeal world, Plotinus uses an analogy from philosophy: "So it is in the case of the soul," he says, "considered as the act of a prior soul, that as long as the prior soul abides, so does the subsequent act" (Enneads IV.5.7). In other words, the soul that proceeded from the One Soul into our body retains its participation in the One Soul (Being in Itself).

This analogy, an eye-glazing abstract scenario to us no doubt, is immediate for Plotinus' students who experience metaphysical reality as self-evident. In Plotinus' turn to metaphysics, he thus explains natural, sensible light as he experiences it using the analogy of a soul's movement as if the intelligible world were more familiar than the physical.[47] He does not start as we might assume, being the most important Neoplatonic philosopher of the age, with philosophy and use light as a metaphor to explain it. Instead, he starts where the poet does with observations of natural light.

That light is corporeal and incorporeal at the same time Plotinus knows from direct experience. As we're lying in bed just before falling asleep, we enter a strange state of neither being awake nor asleep. And here "in the darkness a gleam leaps from within the eye," Plotinus observes, "we make no effort to see anything," yet a "light flashes before us; or we rub the eye and it sees the light it contains," then explains, "This is sight without the act, but it is the truest seeing, for it sees light whereas its other objects [in daylight] were the lit not the light" (V.5.7). Light is thus not "accidental to something else," Plotinus realizes (IV.5.6). We possess our own light, in fact, that spontaneously appears without the aid of the eye, proof that our soul participates in the One Soul, the Absolute where "vision floods the eyes with light." But just like the familiar flash of light appearing to us in a hynogogic state, this is "not a light showing some other object, the light is itself the vision" (VI.7.36). Lying in bed on the verge of sleep, Plotinus seizes on the image presented to him, the spontaneous flash revealing light's essential character as pure activity. And its being neither a physical nor spiritual entity—the body and the intellect fused in the flash.[48]

Like his beloved Plato, Plotinus accepts that the realm of the Ideas and the world of the senses are distinct and that the Soul is immaterial and immortal. But he does not share Plato's dualistic belief that transcendent light is fundamentally different from natural light. Rather, the Light of the intelligible world is one with the light of the sensible world, which he learned from his experience of the light that flashes in the darkness of his body before sleep. You cannot separate these. "Try all you will to separate the light from the sun," he challenges us, "for ever the light is in the sun" (I.7.1). He brings Plato's Sun and the everyday sun together, neither entity being a metaphor for the other but both equally "real." Thus sunlight is not a symbol but a souvenir of the "real" Light, as it is to Philo and the Gnostics.[49]

Living in the Time of Troubles (235–284 CE), when Imperial rule was breaking down, it was testimony to the intensity of Plotinus' experience of light that he held to his mission: impart a true understanding of light so the world will appear in all its beauty and variety to be supernatural, as it really is. Plotinus had managed to find his way to Rome from Alexandria, where he was born and studied philosophy under Ammonius Saccas, and despite unrest in Rome, open a school. After Plotinus' death, his student Porphyry edited and organized his lecture notes into six books called the *Enneads*. He also wrote a biography of The Master, reporting that when asked by a disciple why Plotinus would not accompany him to the sacrificial rites at the feasts of the New Moon, the Master replied, "the gods should come to me, not I to them."

Because it is not the gods who concern Plotinus and drive his Neoplatonic philosophy, it is light. In his experience, this character's two principal actions are to maintain its identity and assert its presence. These lead Plotinus to conclude that because the Light of the intelligible world, the mind, is one with the light of the sensible world, we have access to "authentic being" where "light runs through light" (I.6.3).[50] Plotinus calls this first action light's "integral omnipresence." That is, light maintains its identity in the transcendent world of Ideas *and* in our material world. He perceives that the act of light pouring itself out from the source, the "Rational-Principle" or absolute Mind, into matter (I.6.3), and its act of revealing the objects of our world cannot be separated. Plotinus knows this because he has observed that light preserves its integral self in our world whether "direct or refracted." Its character remains intact, whether "shining through the air" or being "thrown away from the line of incidence" (VI.4.7) as it is by water (i.e., refracted). Our sight itself depends on light's constancy, for our "vision is light or rather," as he says, "becomes one with light."

And because light preserves its integrity, "light sees light," which is its second principal activity, "that is to say a thing sees itself" (V.3.8). A remarkable observation, Plotinus is privileged to experience an aspect of light's character that the wall painters intuited and Philo and John intimated but which eluded them all. In its self-consciousness bound up with its omnipresence, Plotinus sees clearly a profound aspect of light's character. That is, we do not have to consciously act to perceive acts of light whether it is the flash inside or outside us. He seizes on this subtle but familiar experience because it reveals that light in its omnipresence or ubiquity asserts itself simply as presence, its immediate gleam or glare. This is perhaps the most obvious of light's traits and so most difficult to see. Fundamentally, Plotinus experiences light as "pure presence" (IV.5.7). Not substance or spirit but at once sensible and intelligible activity. A presence he perceives as accessible to us as well as to itself in seeing itself.

And this act of the "light of Intellect, or rather," as he says, "light itself, pure buoyant, aerial" (VI.9.9), preserving its integrity and asserting its presence is the "natural beginning of our enquiry" (I.6.1). The light that "shows itself" to Plotinus in "material forms," then, gives rise to his entire philosophy, an interpretation of his light-experience that makes its meaning clear to himself. Furthermore, it gives his meaning a foundation, being a philosophy inextricably bound up with light's biography. His enquiry asks, how does our world of the senses (perceptual) relates to our world of mind (conceptual)? The answer given by his encounter with light lies in the journey of the "Soul" ("Rational-Principle") into sensible phenomena from rocks and trees to us, which is none other than the history of light's activity, the "pouring in of light into matter." This is the story of the Mind (Being in Itself) proceeding from the Absolute One, then to the World Soul (all creation), and finally to the human souls (like Plotinus) and to nature, the phenomenal world. And here we have, of course, the biography of light. So as a participant in this story, it is possible for Plotinus to be "wholly true" to his "essential nature," the true meaning of his encounter with light.[51]

But anyone can know this life in our world if we shatter conventional habits of perception and really see light's "integral omnipresence" and enter into its "pure presence." Plotinus experiences grandeur in this "light not narrowed to any circumscribed form." It is "ever unmeasurable as something greater than all measure and more than all quantity," as he says, the true character of light being neither corporeal nor incorporeal. The impact of this encounter on a Plotinus who finds himself cast into a world in the Time of Troubles is extremely personal. Light's character of pure activity that he encounters on earth, which pours from the other country of the One (Being in Itself), is the beginning of his life story as well as of light's. Aware of acting in both worlds, light is his authentic self, flowing from and participating in the Light Mind. Embracing this light in face of the empire's collapsing frontiers, as Plotinus urges his students to do, "you are self-gathered in the purity of your being," he tells them, "nothing now remaining that can shatter that inner unity, nothing from without clinging to the authentic man." Thus, as he testifies in the first book of the *Enneads*, he finds himself "wholly true" to his "essential nature, wholly that only veritable Light" that had "now become very vision" itself (I.6.9).[52]

Augustine's Seducer

Augustine (354–430 CE), though he was the first Christian philosopher, adopted the pagan philosopher Plotinus' quest for the One, for the "veritable light." He even borrowed his language describing the way to this light as "entering into" himself (IV, 8. 6. 1. 1–11). And both give light a history, a biography that assures them of the reality of their interpretation of an encounter with light, one that recapitulates their own biography. But the Christian who believed that his duty was to approach God inhabited a different world from the pagan who believed that the gods should come to him. And the difference between the gods, Absolute and Christian, on which they project the meaning they give their experience of light leads to different conclusions about its character. For Plotinus, this is the oneness of the sun of our world and the Absolute light of the spiritual world (not Plato's sun as a symbol of the Absolute). In contrast, Augustine insists to the end on the great gap in his experience between the light of this world and the Light of God.

A suspicious character is sunlight whose sweetness, as Augustine admits, "seasons the life of those who blindly" go through the world. All of us in his view are earthlings groping our way through the darkness of sin in need of some delight to make it bearable. Yet he cannot trust his delight. Augustine is tempted, as he admits in his *Confessions*, by light's beauty with its "seductive and dangerous sweetness" (Book X. xxxiv [52]).[53] After Lucretius' excitement about the swerve of light (lightning's flash) and Philo's reveling in the torrent of light, after John's and Hadrian's constructing with contagious enthusiasm their monumental systems of light, and after Plotinus' embracing the beauty of light and devoting his life to our doing the same, we're struck by Augustine's resistance. Each of the refractors we've heard from welcomes acts of light into their lives and the world. In fact, each is happily seduced by sunlight and elaborates a testimony to persuade us of its power, finding the language, whether of science or philosophy or religion to describe its uniquely personal impact.

So we're startled to find Augustine, the first great Christian thinker, resisting light's seduction. A "delight to my eyes," he admits, but prays that its "glowing" and its colors "get no hold upon my soul" (X. xxxiv [51]). He is suspicious of sunlight precisely because it is so seductive, warning throughout his works of its seductiveness. His shrill insistence on its danger, in fact, attests just how powerful light is in his experience, no less so than for Plotinus. It is a light that "makes its way with such power" in the world, Augustine confesses, that "if suddenly it is withdrawn, it is sought for with longing," which Plotinus knew, "and if it is long absent," it has a "depressing effect on the mind" (X. xxxiv [51]), a familiar experience for many. Beautiful and powerful, yes, but this light is a nonetheless subversive character tempting us to savor the sensuousness of the natural "light of every day," leading us away from the "immutable light," the true light that is "utterly different from all our kinds of light" in this world (VII. x [16]). Thus for Augustine there is a great gulf fixed between eternal and temporal light.

Born to a Christian mother and pagan father, Augustine's life spanned the years from the twilight of paganism and the Roman Empire to the noontime of Christianity. As a teenager, he had experienced the Emperor Julian's restoration of paganism in 362 CE, in middle age the resurgence of Christianity under Constantine's successors, and as an old man the triumph of Imperial Christianity. He lived to see an "orthodox" church established, erasing the varieties of Christianity, such as most Gnostic sects and the Messalians whose aim was ecstatic union with the light-body of Christ, that flourished between the letters of St. Paul (c. 50–60 CE) and the sack of Rome by Visigoths in 410, twenty years before Augustine's death.[54]

Throughout his life, Augustine was afflicted by an acute sense of sin and guilt intensified by his sensual nature, which was extremely sensitive to music and light. The omnipresence of a seductive light stirred up his guilt, making him suspicious of the senses. He despaired of the day "which bathes with light all that we see," a light that snuck up on him when he least expected it. And then "comes down upon me with gentle subtlety through many media," he lamented, "while I am doing something else and not noticing it" (X. xxxiv [51]). To alleviate this anguish, he sought the embrace of a community, first in a "religion of light," a Gnostic Manichaean cell, where as a teenager he became a "hearer" and remained for nine years, then in a Neoplatonic academy of scholars who revered Plotinus.[55] But in the end he returned to his mother's religion, joining a Christian community and at the age of forty-two and becoming Bishop of Hippo, a north African port city.

This journey, leading Augustine through Manichaeanism, which was regarded in Carthage as a radical reformed Christianity, and Neoplatonism, especially Plotinus whom he had been reading just before his conversion, was a quest for the true light. Augustine departs from Mani the founder in Iran of Manichaeansim (third century CE) and from Plotinus in his insistence on the absolute separation of the corporeal, "radiant light of this physical world," a seducer, from the "immutable light" of the other world. An intimate conversation with his dying mother leads him to the "conclusion that the pleasure of the bodily senses, however delightful in the radiant light of this physical world," as he says, is "seen by comparison with the life of eternity to be not even worth considering" (IX. x [24]). Ordinary light presents its body to Augustine as seductively sensual. Thus by definition it cannot be part of or participate in the true light he encounters within himself, which as he testifies "transcended my mind," the presence of the eternal God. The "earthly sun" that the "eyes of flesh" see can have nothing to do with the eternal unchangeable light; it is completely "other" than the light that is "Truth," which as he testifies, "made me" (VII. x [16]).

So he is adamant that sunlight cannot be an emanation of the Sun, the Absolute light, rejecting a connection dear to Manichaeanism and Neoplatonism. Augustine abhors light's act of enticement because of its power to dissolve the boundary between the temporal and eternal, flesh and soul. And he dedicates his life to ensuring with all his considerable rhetorical powers that this boundary be preserved. Ordinary light, then, is again a metaphor rather than an actual souvenir of the metaphysical light.[56] Yet Augustine remains conflicted because he achieves his vision of God "by internal reflection and dialogue and wonder" at being itself, the seductive beauty of the "corporeal" world. At the same time, he gains his vision of truth via his love for the "eternal being itself" as he enters into his "own mind" (IX. x [24]). He may, as he says, resist the "pleasures of the eyes of my flesh" that light offers, but he cannot resist its power to act within him (X. xxxiv [51]).

Seeing that everyday light is external yet intuiting that it is internal, he knows that ordinary sunlight and absolute light are not so easily divided. Thus he can't escape conflict and guilt. His experience of sunlight tells him that the physical and spiritual cannot be separated, which is why he resists ordinary light to such a degree, but his will insists on the boundary between them. At the heart of Augustine's experience of light is the absolute chasm he is constrained to keep open, perhaps because of the acute sense of guilt at his own sinful nature, between the unchangeable light of God and the natural light of our ordinary experience. With his conversion to Christianity he may have found salvation, but his conversation with light continued to be conflicted.

Augustine attempts to reconcile this conflict by going back to light's birth. Refracting Genesis, as the Manichaeans habitually did, he interprets the "literal" meaning of light in the first chapter (verses 3–4) as being neither a material nor metaphorical but a spiritual light encompassing our ordinary temporal and the true eternal light. This light not only "made him," he confesses, but all of creation in a series of acts that began with "coming to the knowledge of itself," just as he had to do in order to achieve vision (*City of God*, bk. 8, chap. 7). Thus the light of his experience interpreted correctly reveals a history of light that operated in the world on his behalf. Understood rightly, the ordinary light that comes down upon him with sweetness, catching him unawares, is not a seducer but an agent of knowledge.

Here the biography of light and his own merge. This interpretation derives from his grounding in the "true" eternal light of the Manichaean belief that we all possess a "Light Mind." For Mani, Jesus reveals himself as just such a possessor, declaring, "I am your light, . . . your great mind," paralleling the "Buddha Pure Light" of Augustine's contemporaries in India and China. He associates this Mind with the Holy Spirit. This allows him to ease the guilt arising from his internal conflict: willing a split between ordinary and absolute light but experiencing their unity. He interprets his intense experience of sunlight's character as a spiritual presence, then, albeit physically seductive, with the act of the Holy Spirit in effecting a spiritual oneness within a community of believers.[57]

Augustine died on August 28, 430 CE, amid a burning fever and city. Vandals had invaded, pillaging and setting fire to Hippo as he prayed for two things: the preservation of his flock of congregants and his library of works. Most of his flock managed to flee to safety, and the heroics of friends succeeded in rescuing his library. Augustine's works, which proved to be immensely influential, including their moving testimony of his encounter with light, were miraculously preserved. By these he proved to be with St. Paul, the principal molder of Christian attitudes to God, man, and nature through the twentieth century.

But Augustine's suspicion of light the seducer, though it cast a shadow over the age to follow, could not dim the light to which the Refractors who preceded him testified. His body belonged to the Roman culture in its intuition of light's power to dissolve boundaries, such as between the physical and the spiritual, as it did for Plotinus; yet his will belonged to the Christian church in its determination to separate these. So the glorious light always on the move erasing such lines, which was at the heart of the Roman experience, necessarily threatened him. Fundamentally, the Refractors of the pagan world trusted their experiences of light to reveal themselves to themselves, as they discover their "authentic being," for which they found expression and grounding in symbolic systems and what they imagined to be literal histories of light. These provided solid foundation for the meanings they imparted to the light-experience. Not only does this discovery reveal themselves to themselves, as it did Plotinus, but the world to itself as it becomes illuminated by the sun, as when he says, "a thing sees itself."

The Refractors caught light in the act of being light. They brought light directly into our consciousness by depicting light itself, giving it a portrait, and appropriating its biography. The mosaic artists make us acutely aware of light as light, as we have not been in previous ages, viewers self-consciously witnessing acts of light in the world. The wall painters press this further, presenting us with a depicted light in their work braided with actual light, and the architect of the Pantheon confronts us with this light as a palpable body. The bold figure of light bares itself to the Romans, then, allowing them to discover a new light in human experience. Aware of light as light, they revel in the paradoxes of its character revealed by fresh observations: it reveals the world as it hides itself (Philo); differences in the world while erasing them (John); corporeal objects even as it is not an object, being incorporeal (Plotinus); and itself as external even as it is existentially internal (Augustine). These paradoxes render the Refractors self-conscious in the presence of light's character, which they discover to be an attribute of light itself; for as Plotinus observed, "light sees light."

Roman artists and writers all share this basic encounter with light as light, a fundamentally active and self-conscious light. That is, light does not signify something else because "light is itself the vision," Plotinus concludes. The Refractors discover, as Philo says, that light is self-sufficient in knowing itself and acts autonomously in the world even as we are complicit with it. Such a revelation drives Philo to probe light's history. And so he gives light a biography, versions of which he and John as well as the pagan Plotinus and Christian Augustine alike appropriate. Each claims the story of light coming to be within our world as the working out of light within them to effect vision or salvation or revelation of the true self. And each of these acts is a variant of light's fundamental act in the world—dissolving boundaries—which Plotinus embraced and Augustine resisted.

That this is its essential nature all Refractors agree as they shift from the Greek contemplation of a largely external, static light to their appropriation of a potentially internal, active light. But the specific acts of light at the core of their individual experiences differ. For the artists, the character of light they knew breaks down the threshold between what we assume to be external and internal to us, and for the architect that between materiality and immateriality (Pantheon). Acts of light in the writers' experience dissolve boundaries between the determinate and indeterminate (Lucretius), God and man (John), man and man (Hadrian), and the corporeal and incorporeal (Plotinus). Always on the move, light renders all manner of categories ambiguous, ignoring inherited thresholds.

Furthermore, cooperating with us as viewers, it makes us complicit in its radical activity, which compelled the Refractors to "think about the light," as Poimandres advises. And they did: the distant jagged light arresting Lucretius, and the existential torrent of rays pouring into Philo; the luminous personage appearing to John, and the Heliophany sustaining Hadrian; the inner flash of "pure presence" exciting Plotinus, and the "beauties" of light sneaking up on Augustine. Unlike the Greeks who were preoccupied with light in relation to shadow, the Romans were clearly obsessed with light in relation to themselves. Light revealed something of its psychology in the Greek experience and of its biography in the Roman, the latter finding meaning in projecting its history as their own. Cooperating with this character, an autonomous light that sees itself, ironically, they come to "understand it" in new ways.

Plotinus of the Pure Light, a prime discovery of the Classical Age, and Jesus of the True Light, the engine of the Middle Ages to follow, strangely, met in Constantine's Imperial Christianity that commonly consolidated its status by equating the wisdom of Jesus and Plotinus. Much of Christianity's ultimate success was owing to the intellectual respectability gained by such a strategy, but the power of Augustine's light-experience testimony clinched its triumph and challenged this continuity between the Christian and Classical worlds, so important to Constantine.[58] Although Christianity inherited from Plato the yearning for an ultimate reality beyond this earth and the conception of God as the Good, Christians broke with the Greco-Roman world in substituting for the gulf between the intelligible and sensible realms, which dominated Greek philosophy, the Christian gulf between Creator and creature. Plotinus rejected the former split by recapitulating the biography of light, whereas Augustine reinforced the latter by embracing an alternative history of light as the ground of his personal biography of salvation. In doing so he effected a final break with the pagan world.

NOTES

1. Horace's observation, *Epistles*, II.1.156. During this period, we see a shift from the Athenian spectator who participates as a member of the audience in the theatre to the Roman viewer who interprets as an individual, seeing and reading meaning on walls and in books, a change in the status of viewing (Goldhill, "Refracting Classical Vision," 19, 22–23).

2. Stewart, *Classical Greece and the Birth of Western Art*, 291. After undergoing Hellenization, Roman art per se begins as the Romans conquered Syracuse (211 BCE) and triumphed over Perseus, the last king of Macedon (168 BCE).

3. Henig, *A Handbook of Roman Art*, 117; "skin," Mazzoleni, "Architecture and Illusionistic Painting in the Roman House," 24; "optical light," Summers, *Vision, Reflection, and Desire in Western Painting*, 38.

4. Summers, *Vision, Reflection, and Desire in Western Painting*, 38, 40–41. Following David Summers, I make no distinction between "illumination" or highlights and "reflection"; as he says, "highlights in fact *are* reflections" (37). However, E. H. Gombrich observes that Pliny uses two different words for these: "lumen" for light per se, and "splendor" for gleam or luster ("The Heritage of Apelles," 6). The former reveals surface, he maintains, the latter texture.

5. Dunbabin, *Mosaics of the Greek and Roman World* , 7; Wheeler, *Roman Art and Architecture*, 205.

6. Ling, *Roman Painting*, 218; Gombrich, "The Heritage of Apelles," 6; Summers, *Vision, Reflection, and Desire in Western Painting*, 41.

7. Villard, "Light and Colour (280-150)," 142; Charbonneaux, *Hellenistic Art* , 141.

8. Zajonc, *Catching the Light*, 23–24.

9. The philosopher Richard Rorty suggests three ways in which we assimilate a new experience, reweaving the "fabric of our beliefs and desires": by perception (observation), inference (conclusions drawn from observation), and metaphor. But these do not operate in the same arena. As he says, "both perception and inference leave our language, our way of dividing up the realm of possibility, unchanged." This is the arena of the image. By contrast, metaphor assumes that language, logical space, and the realm of possibility are "open-ended." Metaphor is nothing short of a "call to change one's language and one's life, rather than a proposal about how to systematize them" (Rorty, "Philosophy as Science, As Metaphor, and As Politics," 214). Yet metaphor made the Refractors uneasy because they assumed, as we do, that any object available to their senses, which light announces, offers certain knowledge, but suspected that it may move the object beyond the range of the senses. And does it then pass beyond certain knowledge? The answer is "no" because the boundaries between what is literal, the sun being the prime model, and what is metaphorical break down. The sun is literal and metaphorical. The image, in fact, carries within itself the metaphorical because as Plotinus realized we cannot apprehend or construct an image of the sensible or "real" apart from metaphor. Thus we can have certain knowledge. We'll find the Jew Philo as well as the Evangelist John and the Christian Augustine all caught up in this paradox. We've seen this before, of course, with Plato's sun that dazzles as it blinds because, as Aristotle realizes, there is no escaping "metaphor in the description of metaphor."

The question of whether metaphor moved objects beyond knowledge nagged Aristotle. "Every object of sensation," such as the sun, he says, "when it passes outside the range of sensation, becomes obscure; for it is not clear whether it still exists," because we can comprehend it only through our senses (Aristotle, quoted by Derrida, "White Mythology," 51). Thus the metaphorical as it "carries itself away" in further "deviations" that have no necessary limit, Jacque Derrida asserts, "endlessly constructs its own destruction" (Derrida, 71). This is because it is inherently unstable owing to the fact that there is no reality apart from it. It's metaphor all the way down.

10. Deacon, "The Aesthetic Faculty," 36; Kearney, *Poetics of Imagining* , 142, 152.

11. "Who first instituted . . . one might want," "also began the practice . . . minimal expense," Pliny, *NH* XXXV, quoted by Pollitt (*The Art of Rome c. 753 B.C.-337 A.D.: Sources and Documents*, 116). The painter achieved "unique brilliance" by laying a thin foundation layer for his fresco of fine plaster that takes the paint while still wet, bonding with it and becoming part of the wall. As it dries the colors reveal a special luminousness that he enhances with a mirror-like polish (Pappalardo, *The Splendor of Roman Wall Painting* , 9; Zanker, *Roman Art*, 11, 26–27).

12. Only glimpsed in the background of Greek art, landscape now emerges in the foreground. Villard, "Light and Colour (280-150)," 142, 167, and "Landscape, Nature and Realism," 172; Ling, *Roman Painting*, 142; "engages in self-reflection," Donald, "Art and Cognitive Evolution," 19.

13. In the Villa Oplontis (first century BCE), for example, which may have been owned by Nero, the rays of the winter solstice sun at their lowest angle create a "tenuous light" that

probes the painted wall. And those of the summer solstice sun at their highest illuminate the colonnade of the central garden with a strong direct light (Zanker, *Roman Art*, 120; Mazzoleni, "Architecture and Illusionistic Painting in the Roman House," 28).

14. Leach, *The Social Life of Painting in Ancient Rome*, 9–10, 37; Thebert, "Private Life and Domestic Architecture in Roman Africa," 357, 393, 407; Ling, *Roman Painting*, 213–17. Romans understood that art could be viewed in two different ways, described by two later writers who offer specific instruction. In his *Imagines* (third century CE), Philostratus the Elder instructs a boy whom he is taking on a gallery tour to enter a painting completely, even to the point where he can "smell" it in order to really understand its images. We trust a guide who observes, for example, that "the moon sheds a light that the eyes cannot quite trust" and calls attention to the "gleaming" in a still-life bowl of milk (Philostratus, *Imagines*, 2.29, "Antigone," trans. Arthur Fairbanks, Loeb Classical Library). And this intimate relationship is necessary to sharing the experience of light embodied in wall paintings that constitutes the painted image itself. We view painting not so much to decipher it, then, as to experience its numinous presence (Francis, "Visual and Verbal Representation," 291, 296–97). By contrast, Cebes of Thebes in his *Tabula* (end of the first century CE), teaches an allegorical approach like that used by Ovid advising a young, procession onlooker on how to impress his girlfriend (*The Art of Love* , Book I, Elegy V, trans. A. S. Kline, 2001). Cebes views the picture as a door to a spiritual reality beyond the viewer's ordinary reality, thereby initiating him into truth. For Philostratus, pictures represent the world; for Cebes, they are a bridge out of the world (Elsner, *Art and the Roman Viewer* , 22, 88). Clearly, for the more personal expressions captured by wall paintings, Philostratus' approach is the appropriate one.

15. In their country houses, villa owners who admired Greek culture often displayed Greek-influenced art, which in the city could offend local Roman sensibilities, but this art employed a specifically Roman visual language (Zanker, *Roman Art*, 10). Stage scenery of the Hellenistic and early Roman theatre, Peters maintains, influenced the development of landscape painting (Peters, *Landscape in Romano-Campanian Mural Painting*, 67). The cluster of buildings is on the west wall of the antechamber, "Boscoreale: Frescoes from the Villa of P. Fannius Synistor," an installation, bedroom O, New York Metropolitan Museum of Art.

16. Wheeler, *Roman Art and Architecture*, 196; Dorigo, *Late Roman Painting* , 38; "quivering haze of broken gleams," Maiuri, *Roman Painting*, 120–21.

17. "Outside" and "inside," Mazzoleni, "Architecture and Illusionistic Painting in the Roman House," 25; David Grandy, "The Otherness of Light: Einstein and Levinas" (*Postmodern Culture* 12.1 [September 2001]), section 36.

18. "Fabric," Bryson, *Looking at the Overlooked* , 45, 51; Leach, *The Social Life of Painting in Ancient Rome and on the Bay of Naples*, 125. Norman Bryson concludes that in all four of the "Pompeian styles" of wall painting, as August Mau designated them in 1882, "what is actual (the experience of local interior space) is invaded by a principle of irrealisation or fiction" (*Looking at the Overlooked*, 35). They have a shared aspiration to "negate the physical limits" of the architectural rooms, which were "cramped" and "compressed," in order to go past the "parameters of the real and allow actual space to be penetrated by a fictional expanse beyond it" (45). The First Style (200–100 BCE) creates an "intermediary zone" between "reality and representation," its "peculiar achievement" (36). The Second (100–15 BCE) goes all out in creating a "simulated space" that takes over from the "real space" of a villa's rooms (37). The Third (15 BCE–50 CE) highlights "*transitions* [italics his] between reality and simulation," whereas the Fourth (after 50 CE) revels in the "breaking open of the wall and the expansion into fictive space" (43), so the actual room becomes a theatre set. Reality is "playfully invaded by its fictions; the room is gathered and absorbed into the space of simulation" (45). All styles show the influence of the scenic backdrops used in Roman theatre (Leach, *The Social Life of Painting in Ancient Rome and on the Bay of Naples*, 94, 97).

19. Pappalardo, *The Splendor of Roman Wall Painting* , 104; "dematerializes," Mazzoleni, "Architecture and Illusionistic Painting in the Roman House," 32.

20. Hanfmann, "Hellenistic Art," 140.

21. Bryson, *Looking at the Overlooked* , 59.

22. Grandy, "The Otherness of Light," sections 6, 47.

23. Esolen, "Introduction," 2; "sit calmly writing . . . hour of trial," Lucretius, *On the Nature of Things*, I. 41–42; "tough-minded grandeur," Rexroth, *Classics Revisited*, 60. All quotes from *On the Nature of Things* (*De Rerum Natura*) are from Esolen's translation, unless otherwise specified.

24. Serres, *The Birth of Physics*, 7.

25. "Realized in nature," Serres, *The Birth of Physics*, 7.

26. "Naturalistic conception of things . . . mythology," Santayana, *Three Philosophical Poets*, 21; Serres, *The Birth of Physics*, 31, 112; Minadeo, *The Lyre of Science*, 11. Michel Serres calls Lucretius' poem a "discourse on physics." Interpreting his experience of light, that is, Lucretius employs a powerful metaphor, which becomes for him a model of the universe. In his poem's opening we are to choose between two kinds of physics: indeterminate and determinate. The former symbolized by Venus (eros) includes chance, and the latter symbolized by Mars (death) does not, being committed to destiny and fate. The poem chooses Venus, refusing to complete the circle from the opening of the poem to its closing, from birth to death, and choosing instead a "spiral"; so whirling turbulences or "vortices" appear everywhere in the poem, just as "stochastically in space and time, turbulences appear." This "model is of such fecundity for the physical and natural sciences that it is at the horizon of four centuries of research" (*The Birth of Physics*, 35, 126–27).

27. "Turns endlessly back upon itself," Serres, *The Birth of Physics*, 117. The unpredictable, impersonal nature of his wild light is why Lucretius evades traditional associations of darkness and light, taunting us with contradictory images even as he comforts us by alleviating our psychological torment, our fear of death. He cycles his dominant images of light and dark to the rhythm of the creation/destruction design of his poem, creating an "irrational poetic disturbance" (West, *The Imagery and Poetry of Lucretius*, 83). This arises from his viewing light in fluid terms, impossible to pin down as it "irrigates the sky," the sun being a "fountain" (V.281–2, 290–294), but then becomes an ever-shifting fabric (V.136–49). No wonder at times he subverts traditional light symbolism outright, declaring that "this night of the mind must be dispelled / not by the rays of the sun, nor day's bright spears, / but by the face of nature and her laws" (III.91–93). In other words, forget the clichéd associations, and observe nature. But he then does an about face and confesses, "I turn the bright light of my verse / on darkness" (IV.8–9).

28. Siegert, "Philo and the New Testament," 195–96. In the distinction between "writer" and "author," I follow Raymond E. Brown in his edition of *The Gospel According to John*, who reminds us that the attribution of the Gospel's authorship is problematic and distinguishes between its "author" whose ideas the book expresses and its "writer" who collaborates, records, and edits the traditional materials (I. LXXXVII, XCIX–C). The light that John apprehends as a personage is an existential experience of the true nature of our being (Pulver, "The Experience of Light in the Gospel of John," 252). The fullness of the day's light in John's experience creates the identity Life-Light, which Akhenaten apprehended before him. But John goes further. Light's plenitude pervades his body. He extends this image of Life-Light to encompass the identity Logos-God. Though this may seem an abstraction to us, for John it is not. "Logos" to Jewish readers would have conjured up the original creative act of God, "God said, Let there be light; and there was light" (Gen 1:3). It is by the "word of the Lord" (Psa 33:6) that all is created and holds together. Greek readers would have understood Logos as the active rational element informing reality, including us (matter being the passive element).

29. "One does not say . . . the second is only a figure of speech," quoted by David Park, *The Fire within the Eye*, 91–92. On John's vacillation, see *Theological Dictionary of the New Testament*, 1295. On John's use, see Petersen, *The Gospel of John and the Sociology of Light*, 13. Max Pulver puts this in a Gnostic context, saying that what is expressed in John's Prologue hymn is "light, not as a metaphor, . . . not as a symbolical or even metaphorical expression for the ineffable. These words refer to light in its psychic function, as the inner process of illumination and as the psychic content of illumination; they refer to light as a path, in the sense that the Tao is a path. This light gives insight, inner sight; it is illumination and wisdom in the Gnostic sense" ("The Experience of Light in the Gospel of John," 241). Literal as a psychic reality, but John points to an internal rather than external referent here. Pulver underplays his encounter with the fullness of brightness.

30. Word as light before sound, Silvestri, "From the Eloquence of Light to the Splendor of the Word," 118. Primordial light in present, Borgen, *Logos was the True Light*, 104. For John's rewriting of Genesis, see Carmichael, *The Story of Creation*, 33–35.

31. Norman R. Petersen, *The Gospel of John and the Sociology of Light*, identifies three pairs of conceptual systems that John employs, revolving around the Word, the Bread, and the Light. Each is independent of but maps onto the others. The Light system is primary, however, and "nonmetaphorical." It alone has a "literal form of reference to the Other during the period of the incarnation—of the Word which was the Light" (63–64).

32. "Differentiation . . . nondifferentiation" and "luminosity of the Word/Light," Petersen, *The Gospel of John and the Sociology of Light*, 73, 75. In the Manichaean tradition, as Max Pulver points out, the "luminous self" that becomes identified with the "luminous soul" is often called Jesus, though John does not seem to be influenced by this ("The Experience of Light in the Gospel of John," 242–43).

33. On Hadrian's relation to the architect, see MacDonald, *The Pantheon*, 12, 67; Henig, *A Handbook of Roman Art*, 38–39; "totally suffused and inhabited by light," Mark Major, et al, *Made of Light: The Art of Light and Architecture* (Basel, Switzerland: Birkhäuser, 2005), 26.

34. Fabric of brightness, Wheeler, *Roman Art and Architecture*, 13; "made for light," MacDonald, *The Architecture of the Roman Empire*, I, 118. Hadrian may have borrowed this idea of an opening in the dome to admit light from the Emperor Nero (37–68 CE). His extravagant "Golden House" (*Domus Aurea*), whose vestibule was outfitted with an ivory ceiling that could rain down flowers and pipes, which sprinkled guests with perfume, had an octagonal hall with an oculus (Pappalardo, *The Splendor of Roman Wall Painting*, 160). But Hadrian shrunk the oculus and placed it higher, positioning it carefully in relation to the interior space, which gave it prominence and flexibility.

35. MacDonald, *The Pantheon*, 74–75.

36. "Great shaft . . . solar effusion," MacDonald, *The Architecture of the Roman Empire*, I. 117–18; "perfect unity of the light," Lucae, "On the Meaning and Power of Space in Architecture," I, 559.

37. "Unify unities," MacDonald, *The Pantheon*, 88, 90; "order, peace, harmony, unity," MacDonald, *The Architecture of the Roman Empire*, I, 121. With the new world arriving, the forces of traditional Roman religion weaken as other religions like Mithraism and Gnosticism gain in popularity. Syncretism, which his travels impressed upon him, dilutes traditional religion. The strict contract between the servant, worshipper, and lord was breaking down.

38. Identify with Phoebus, MacDonald, *The Pantheon*, 89. Hadrian, in contrast to Nero, substituted for his own image that of Zeus-Jupiter-Helios. The Romans equated Helios with Sol who was both a "cosmic metaphor" and a "fully fledged Roman deity," as S. E. Hijmans reminds us, *Sol: The Sun in the Art and Religions of Rome*, 483. Sol was a god with "deep roots and a modest yet clear presence in the city's religious landscape and calendar." Its cult, Hijmans concludes, flourished in Rome continuously from the Republic through the first centuries of Empire and beyond, but did not involve a "congregation of sun worshippers." Sol was not restricted to any group because the "godhead of Sol was not a matter of faith but of knowledge, primarily astronomical." The public cult was not such that "one chose to adhere, or not, but was a matter of observing proper ritual (sanctified by venerable traditions) in order to maintain correct relations with this particular divine entity." Significantly, given Hadrian's intentions, the Romans "did not differentiate between the physical sun and the sun god" (507–08).

39. "Split a piece of wood, and I am there," Barnstone, *The Other Bible*, 305. *The Gospel of Thomas* could well have been composed about the latter part of the first century CE. See Ron Cameron's entry in *The Anchor Bible Dictionary* (VI: 536), Marvin Meyers' edition of *The Gospel of Thomas*, 10, and Elaine Pagels' introduction to *Beyond Belief: The Secret Gospel of Thomas*, 34. Pagels compares The Gospel of John and *Thomas* at length (30–73).

40. The contrast is stark in their respective identities of light. Jesus says, "I am the light of the world" (8:12); Thomas's Jesus declares instead that "There is light within a man of light, and he lights up the whole world" (Barnstone, *The Other Bible*, 302). John in his Gospel strives to confront us with his personal experience of light, which will lead us to believe that he is who he says he is, whereas Thomas seeks to show us the way to the light, which will lead us to the knowledge of the light within us.

41. "We came from the light, . . . manifest through their image" (saying 50), Barnstone, *The Other Bible*, 303; "luminous body," Pulver, "The Experience of Light in the Gospel of John," 244.

42. Barnstone, *The Other Bible*, 569–70. The title of the treatise may have meant in Coptic "knowledge of the Sun-God Ra." Traditionally said to be by Hermes Trismegistus who was aware of Judaism and influenced by Egyptian Hellenistic Platonism, even employing Socratic dialogue as does John in his Gospel, the *Poimandres* had its origins in Egypt in the first century BCE. And Poimandres himself is the shepherd of men, which may have influenced Jesus' identifying himself as the shepherd in John's Gospel (Jn 10:14). Hermes Trismegistus is, of course, more a tradition than a single author (Barnstone, *The Other Bible*, 568).

43. "Father of all, . . . like to Himself" and "all things are filled with light," quoted by Pulver, "The Experience of Light in the Gospel of John," 245. Various Gnostic sects relate variations of light's biography, many elaborated in great imaginative detail as A. P. Smith summarizes (*The Gnostics*, 13–33), and of the experiences of this true light that is true humanity. At the center of the Gnostic imagination is an overriding and urgent concern for the "fate of the divine spark" in us here in the world, our true Mind, where it forgets its origins in the Light world. And it yearns, therefore, to return to this world, being liberated from our body, the material world of shadows, to become the "same substance as the divine world, primordial, archetypal light" (Filoramo, *A History of Gnosticism* , 38–41, 47). Gnosis as a practice is "essentially the experience of light." Although Max Pulver characterizes this light as "the inner light, . . . an inner illumination and radiance created by God" ("The Experience of Light in the Gospel of John," 250), for Hermes encountering Light, his true Mind/Life (Poimandres), the distinction between "outer" and "inner" light is meaningless. Pulver recognizes this, admitting that when John in his Gospel and Hermes in the *Corpus Hermeticum* "speak of light, they do not . . . mean a mere subsidiary force or a particle in the pleroma," the world beyond the sensory world, "but a crucial and concrete inner experience. It is *not a fantasy or an adventure, but an experience*, the essential experience of the true essence" (252; italics mine).

44. "Noon-day beam . . . torrent," Philo, commentary on Genesis XXIII (trans. C. D. Yonge); "best of all existing things," quoted by Chidester, *Word and Light* , 33.

45. "Unmingled and pure light," Philo, commentary on Genesis XXIII (trans. C. D. Yonge); "full of every description of light," Philo on Genesis XXV; "invisible . . . God, being his own light, is perceived by himself alone," quoted by Hillar, "Philo of Alexandria"; for comparison of Philo and Plato, see Chadwick, "Philo," 139.

46. "Light-Stream from God," Goodenough, *By Light, Light*, 11, 148. Goodenough traces the Light-Stream tradition and its main types of expression (*By Light, Light*, 12–15). In both Genesis and the *Timaeus*, which Philo particularly cherished and very likely had by memory, this story involves the history of darkness as well (Blumenberg, "Light as a Metaphor for Truth," 42). Plato places them in distinct categories, whereas Philo gives them different histories. In his refraction of Genesis, darkness is born first. At the birth of the pure light, which is "perceptible only to the intellect," they clashed like brothers, and light's older brother his "adversary darkness yielded." God seizes the opportunity to place a "wall between them." At this moment, the lives of light and darkness part as God separates them, "knowing their opposite characters, and the enmity existing between their natures" (Philo commentary on Genesis IX). Thus are born day and night, the sun and moon.

47. Schroder, "Light and the Active Intellect in Alexander and Plotinus," 248, and *Form and Transformation*, 28. All Plotinus quotations are from *The Enneads* (trans. Stephen McKennna).

48. Aristotle explained that this flash was owing to the shine of the pupil's polished surface when the eye moved rapidly, merely a function of the eye's physiology (proven to be correct, though he misunderstands the mechanism producing the flash). Plotinus, in contrast, emphasizes the "natural character of this internal light" and points out that it is a fact of our experience just as is our perception of external objects. Thus the internal light is "implied empirically in nature itself" (Corrigan, "El simbolismo ," 54). It is "not only an aspect of the physiological mechanism of the eye," but more important, it is the "capacity of the intellect to see itself as pure light." What the eye sees when it sees its "own interior light" is a natural/intelligible light (55), perceived by the senses *and* the intellect.

49. McKenna, trans., Plotinus, "Introduction," 11. For Plotinus, that is, "physical light as such is not substantial" in Plato's sense, as Kevin Corrigan summarizes. "It is only when we grasp the productive power present to it that we can grasp its substantial meaning. If light or sight, then, is the highest of physical activities, and if the activity is an expression of intelligible power, then intelligible light is a natural and philosophically necessary completion of, and source for the understanding of, physical light" ("Light and Metaphor in Plotinus and St. Thomas Aquinas," 193). "While intelligible light is a metaphor, it is simultaneously in the order of substance," for Plotinus a "natural and philosophically necessary completion of, and even ground for, physical light" (197). That is, the physical and the spiritual exist together in the "same logical space" (Corrigan, "El simbolismo ," 55). In the intelligible or spiritual world of Ideas the metaphorical and the literal are interchangeable (Corrigan, "Light and Metaphor in Plotinus and St. Thomas Aquinas ," 199). For Plotinus, "light is a physical-intelligible continuum: movement, quality, form, activity, power, all are implied by its meaning" (Corrigan, "El simbolismo natural ," 52). He thus switches back and forth between using the metaphysical to illuminate (literally) the physical and vice versa. His analogies, which he employs often, imply in some explanations that the activity of the soul is a metaphor for light, in others that the activity of light is metaphor for the soul (e.g., E V.3.12, 17; VI.4.7). Consequently, he is critical of his own use of analogy because no fixed line exists between the metaphorical and substantial (e.g., E V.3.9). Like Jacques Derrida he seems to be aware that "it is not so much that metaphor is in the text of philosophy . . . —rather these texts are in metaphor" ("White Mythology," 60).

Because he adopts the primary Gnostic conviction in contrast to Plato that we can appropriate the fullness of life on earth by the disciplined shattering of perceptual habits and ascent through levels of self-knowledge, Plotinus sees the Gnostics as rivals. But because he views the natural and supernatural worlds as bound up together, he condemns them in *Against the Gnostics* for their contempt of the world. As we see, Plotinus is rooted in his own experience, using sense images with great force and originality (Armstrong, "Plotinus ," 221), demonstrating like a poet his love of the world. Strangely, he ignores the Christians, especially because when he studied under Saccas, a fellow student was Origen (185–254 CE) who became one of the century's most important theologians and a church father.

50. "The gods should come to me, not I to them," O'Meara, *Plotinus* , 3; two principal actions, Armstrong, "Plotinus, " 223.

51. "Pouring in of light," Herman, *The Cave and the Light*, 140–41. This is so because Plotinus' experience of light's presence, its "integral omnipresence," presents him with a paradox: physical light is in reality intelligible light (Corrigan, "El simbolismo natural ," 53).

52. Departing from Plato whose main concern was the ethical, Plotinus embarked on what could be interpreted as a religious quest to achieve this vision, transcending his body and gaining unity with the One (see Frederick Copleston, *A History of Philosophy* , I, part II, 215–16). His encounter with light, however, demonstrates that Plotinus struck out on a primarily intellectual rather than religious journey, guided by the character of light he experienced, though it resulted in his individual salvation as he testifies. Yet his vision could be interpreted as a mystical experience, which Porphyry in his *Life of Plotinus* reports that he had on occasion. Defined broadly as "simply intimate experience of the divine" gained in "whatever way," as K. Corrigan does, we could say that Plotinus is a mystic (*Reading Plotinus* , 30, n. 4). For he says that he has often "woken up to myself out of the body and entered into myself, outside of other things, seeing a beauty of great wonder and trusting that then above all I belonged to the greater part, activating a best life and coming to identity with the divine" (IV, 8. 6. 1. 1–11, quoted by Corrigan, *Reading Plotinus* , 31). This union with the intellect is an entering into himself, but is it a "mystical union with the One," as Corrigan concludes? Not easy to answer, he admits, for there is "no simple dividing line between literature, experience, mystical theology, or philosophy" (*Reading Plotinus* , 31, no. 4 cont'd). And using Mircea Eliade's criteria for mystical experience doesn't help. Contrary to the classic experience of the "Mystic Light," according to Eliade, the "veritable light" that Plotinus embraces does not erase the surrounding world. Plotinus from beginning to end savors the beauty and particularity of the natural world. Nor does it bring this world to "primal perfection," for he is already acquainted with the perfection of the natural light and the presence of the supernatural in the natural (Eliade, *The Two and the One* , 75). Plotinus' journey was through the sensual world to self-knowledge, not

to God (O'Meara, *Plotinus* , 108). A century later, the emperor Julian (331–63 CE), last of the pagans, driven by his "extraordinary longing for the rays" of the sun, takes this path. Like Plotinus he revels in the light of nature, expressing his gratitude in his *Hymn to King Helios*, but its "essential nature" is to operate as mediator, preparing Julian for the work of King Helios to effect his union with the divine (Julian, *The Works*, 131, 133, 159).

I'm reminded of the Emperor Balbinus in Guy Davenport's short story "C. Musonius Rufus" who notices things like the "shiver of light" in a "fidget of leaves" and concludes that it "goes in all directions at once and thus balloons out from its source. Spirit must be a substance like light" (*Da Vinci's Bicycle* [Baltimore, MD: Johns Hopkins University Press, 1979], 10, 12). Decidedly, this is not a mystical experience, and it is at the core of Plotinus' experience. Light is both a spiritual and physical phenomenon, yet for Plotinus there is only one "object" of "true knowledge: light itself" (Blumenberg, "Light as a Metaphor for Truth," 41). He would understand completely the modern philosopher Emmanuel Levinas' remark that "the miracle of light is the essence of thought" (quoted by Grandy, "The Otherness of Light," *Postmodern Culture* 12.1, section 36).

53. All references to Augustine's *Confessions* are from Henry Chadwick's translation unless otherwise noted.

54. Imperial Christianity, Markus, "Augustine," 341; Messalians, Eliade, *The Two and the One* , 61.

55. The Manichaeans were Gnostics. Their teachings achieved the fullest elaboration, reached the greatest number of people, and lasted the longest, almost fifteen hundred years, of any Gnostic sect. Mani (c. 215–75), who was a contemporary of Plotinus, knew the Hebrew and Christian scriptures and modeled himself on St. Paul, considering himself an "apostle of light" and traveling to India and Babylonia as a missionary (Smith, *The Gnostics*, 126). Like Paul (Acts 9:3; 22:6, 9, 11; 26:13, 18), tradition reports that he had a vision of light at twelve years old, seeing the "King of the Gardens of Light," God himself. For Mani god is light, literally not metaphorically, and his abode is light (Park, *The Fire within the Eye*, 24). Mani's universal "religion of light" penetrated to the farthest reaches of the Roman Empire and China whose adherents called Mani the "Buddha of Light" (Barnstone, *The Other Bible*, 670). In some Roman cities Manichaeanism was considered a radical reformed Christianity (Brown, *Through the Eye of the Needle* , 158).Its elaborate mythology detailing light's history was so inventive that an exasperated Augustine said they "dream up everything imaginable" (Barnstone, *The Other Bible*, 675). This history unfolds in three phases: first, eternal light and darkness appear, then darkness attacks the light, which results in a mixture of light and dark, and third, light struggles to wrest light from darkness. To facilitate this, the King of Light creates the material cosmos and our world. Though the light within the human soul is paramount, fragments of light exist in everything, including us. So the world is the King of Light's plan to ultimately rescue these, separating light from darkness, and restore them to his kingdom (Smith, *The Gnostics*, 132, 136–38). This myth was extremely persuasive, its biography of light seizing the imagination of countless people across many cultures, providing joy and solace.

For one of Manichaeanism's psalms proclaims, in "All Life" in "every Place" the "Remnants of Light" exist, which God will gather to himself (Barnstone, *The Other Bible*, 329). The ritual meal of the "hearers" was thus dedicated to the salvation of light itself (Baker-Brian, 125). It sought to liberate the light in the food they ate as well as in their souls. And they facilitated this by practicing chastity and self-denial, which greatly appealed to Augustine. A hymn they sang at this meal yearns for Jesus to appear, shining with radiance and identifying himself, saying, "I am your light, radiant, primeval, / your great mind and complete hope." And as the "Light Mind" he promises, "I shall cleanse from you through perfect light" all the "corrosion that you have passed through" (Barnstone, *The Other Bible*, 322). Looking up at the waxing moon, a "ship of light" filling with liberated particles of light so cleansed, offering hope, inspired countless people (Smith, *The Gnostics*, 139).

56. Blumenberg, "Light as a Metaphor for Truth," 43. Although Augustine recognizes a number of different kinds of "light," such as the "uncreated" light that is God and the light that is truth, his core light-experience pivots on this axis between an "external physical" (or "corporeal") light and internal (or "incorporeal") "psychological light responsible for the soul's sensitive and intellectual powers," as David C. Lindberg notes. These can be reconciled only by the

"intellectual light of divine illumination" (*Roger Bacon and the Origins of Perspectiva in the Middle Ages: A Critical Edition of Bacon's* Perspectiva *with Introduction and Notes.* [New York: Oxford University Press, 1996], xl).

57. "Buddha Pure Light," Barnstone, *The Other Bible*, 322; Brown, *Through the Eye of the Needle*, 159. Yet Augustine ultimately rejected the Manichaean God, a "vast luminous body," condemning as the sin of pride their conviction that because of our inherited inner spark of this light we do not need God's true light. Thus he remains in conflict to the end. For Augustine's contemporary St. Basil of Caesarea (330–379 CE) interpretation of Genesis, employing it in an explanation of time that diverges from Augustine's famous discussion in Book XI of his *Confessions*, as St. Basil posits that "light is the common ground of the whole universe intelligible and sensible," see Arthur M. Young (*The Reflexive Universe: Evolution of Consciousness* [San Francisco: Delacorte Press, 1976], 24). For Augustine's younger contemporary Athanasius' (296–373 CE) view of Christ as radiance, which echoes Mani's in that Jesus shines with the "radiance of eternal light," see Pelikan, *The Light of the World* (30, 40–41, 45, 59, 68). This Christ parallels "Buddha Pure Light" called Amita (or Amida), "rays of light streaming from his pores." Among his other names is the "Buddha of Measureless Light" or "Boundless Light," or "Unimpeded" or "Unopposed" (Yiengpruksawan, "The Eyes of Michinaga," 229–30). The core of the Pure Land teaching, which was brought from India to China in the second century CE, stated that the body of the Buddha radiates beams of light, and, in fact, he even speaks with rays of light. This teaching no doubt influenced the light-mysticism of contemporary Taoism that revered "radiance" above all, though the Taoists concentrated on the sun, supreme luminary. In their magical practice for "absorbing" the "image of the sun," a man stands in the early morning sun holding in his left hand a piece of green paper with the character for the sun inscribed on it and later in his house ingests it (women do the same except they stand in the moonlight with a piece of yellow paper in their right hand) (Needham, *Science and Civilization in China* , v.5, part 5, 83, 182–83).

58. In fact, the main ideas of Christianity in Constantine's view were anticipated by Plotinus, such as the distinction between the physical and the spiritual. We remember that Augustine went directly from reading Plotinus to accepting Jesus and converting. And he makes it clear that Plotinus provided the lens by which the Church and scripture appeared intellectually satisfying. Yet he could not accept Plotinus' blurring the boundary between temporal and eternal.How closely Jesus and Plotinus were formally associated is revealed by contemporary sarcophagi sculpture. In one Jesus is enthroned with a scroll folded in one hand freeing his other to make a gesture of instruction, flanked by Peter and Paul who play the role of the philosopher's muses (c. 359 CE; Harley, "Christianity and the Transformation of Classical Art ," 309, fig 21.1). In a complementary sarcophagus Plotinus is seated and reading from a scroll he holds open in front of him flanked by two muses (late third to fourth century CE; Harley, 313, fig 21.4).

Conclusion

Light is a wonderfully insistent presence in our life. It manifests itself in subtle and not so subtle ways, as we've seen, such as in rays spearing through clouds or the dazzle of noon sun pouring into us. And whether we hold our encounters with it at arm's length or embrace them, we have found ourselves in relationships with a gloriously insinuating character. For rather than something that shines on us, we enter light as light enters us. We may find this exhilarating or disturbing, but we cannot separate ourselves from light or light from ourselves. And our interactions are fundamentally unconscious physical ones, inseparable from our sensing body (photons registering on our eye: light's reflection, refraction, and diffraction). We participate with light in a mutual enterprise. And by exploring this glorious and mysterious reciprocity, another dimension in the history of consciousness is revealed.

Because we cannot escape its enveloping us, cannot step outside light as we host it, we experience its complex "psychology" and obscure "biography" imaginatively, inspired by the endless facets of its character. My story has been that of individual encounters with light. Some witnesses, as we've seen, seizing on its brightness or radiance, others on its rays or glow. And they find themselves in each case living with a palpable presence. Often light being instructor, sometimes judge. It is a fundamentally paradoxical character, because if not for its interacting with matter, refracted by droplets of moisture in the sky or reflected from objects around them, they would not see light. This character invisibly reveals the visible world.

And from the earliest times, we have found ways of making sense of these encounters as well as giving conscious expression to them. Interpreting their core experiences of light, seers to sages have charged them with a multitude of values. They have been compelled to discover the significance of light-experience in their lives and express this, weaving systems of meaning ac-

Figure 5.4. Bay Spotlight. Crepuscular Rays Over San Francisco. ***Bay Spotlight (c) 2012 Dan Peralta-Cieloha***

cording to individual needs and desires. They imagine some experiences of light as revealing the structure of the cosmos or the anatomy of light itself; or providing the key to their salvation or the very foundation of the world and all that is in it, including their own existence. In addition, these meanings derived from their most basic encounters with light often have great cultural impact. Neoplatonism, borne of Plato's epiphany, reigned for centuries and greatly influenced Christianity, a fabric of symbols woven to a large degree from Augustine's fraught encounters with light, to say nothing of John the Evangelist's communion with the figure of light, the Logos-Jesus.

THE WEAVE OF IMAGE AND MEANING

We've seen that these encounters and interactions with light are inseparable from personal testimonies relating the varied experiences of light centered in a palpable "core" image of light. And testimonies we know to be interpretations of what is seen, inferences from our experience. Our leading character, then, proves to be none other than patterns in the mind, our perception and imagination together creating personal meanings. These are driven, in accordance with our visual experience, by metaphors, as we've seen, such as "gnomen" and "line." Both being metaphors for a reality beyond our senses, respectively, the ratio of light/shadow as the way of worlds beyond ours, leading to Thales' cosmic imagination of light, and the line or arrow as a ray, resulting in Euclid's "geometric" imagination of light.

To the shaman, similarly, pigment was a metaphor for the body of light just as to the seer, quartz was for the power of light, though neither recognized a distinction between literal and figurative. Millennia later, pigment for the Roman wall painters was metaphoric as well. But they were self-conscious about the distinction and wrestled with it, well aware that paint was and was not light, literally being light and figuratively representing light. As did Plato wrestle, whose core-experience of light was bound up with a metaphoric sun. For Plotinus, the metaphor of light "seeing," a personification, drives the interpretation of his experience. Light unbidden in the dark, flashing within his eye, forces him to conclude that "light sees light." Investing these physical light encounters with personal values, such men drew threads from their psychological and biographical history and wove larger patterns of meaning. Moreover, in their desire to further ground this tapestry, establishing its existential significance, they often turned to symbols and beyond to symbolic systems, commonly metaphysical or religious.

We've seen how our core-experience of light—the initial image in an encounter that lodges in our consciousness—easily takes on metaphoric meaning by our imaginative weaving. This experience was commonly charged with religious and aesthetic valuations, often overlapping. And following the prehistoric eras, with the rise of the theoretic mindset, these "glorious" and "mysterious" aspects of light, which Reverend Hunter identified, take primacy over its "common" and "simple" aspects. Because light is beautiful in itself, seeming to rejoice in its own beauty, it becomes the ground of esthetic value. Likewise, because our encounter with light can feel like liberation from a "profane universe" or turbulent times, opening a gateway to the "transcendent and the holy," it often becomes the foundation of religious conviction. This explains why light is virtually a universal religious symbol and has been the "preferred key in the music that is mysticism."[1] But we've alluded to the fact that as light's beauty or mystery is elaborated further, this can alienate us from the manifold character of light. The fabric of our light-experience can be torn, the imaginative interpretation of light nearly severed from the core image. And as imagined, symbolic systems of light's personal meaning become culturally mainstream in artistic or religious traditions, they often congeal into conventional forms over time.

Yet neither the shamans and seers, nor the priest-kings or luminaries and refractors we've met ripped the woof threads of interpretation from its warp cords of image, from the core-experience, their life-changing encounter with light. Often this weave of image and interpretation in the human consciousness, we've learned, resides in light's paradoxical character or the need to resolve personal contradictions. The Luminaries came face to face with this character in light's contrary relationship with shadow, to say nothing of its paradoxical invitation to "look for the beauty of light in its invisibility."[2] And the Refractors dealt with tensions triggered by contradictory encounters with light, such as its being man *and* god, or seducer *and* savior.

When this fabric is torn, when interpretation parts from the fundamental image given by a light-experience, conceptual isolated from sensual, the unique character of light in a given experience is abandoned. Aware of this, Akhenaten rejects the traditional gods to ensure that he does not lose sight of the energetic and generative light of his core-experience, which the earlier priest-kings had done. Later, Plotinus weaves a metaphysics from the beauty of light generated by the arresting image of a hynogogic flash, but does not cut himself off from his physical encounter with a glorious light, while succeeding Neoplatonists do. Likewise, Augustine weaves a theology from the mystery of light, its promise combined with threat, but retains bonds to his initial experience of being waylaid by a seducer light, its beauty being so overwhelming as to be dangerous. The Christianity that later congealed, of course, largely severed these bonds, reifying light as abstract religious symbol, divine light being the highest spiritual attainment (a development common to many religions, such as Zoroastrianism's "principles of light," Buddhism's "pure light," Islam's "light to walk in").

THE STRANGE EPISODIC HISTORY OF LIGHT

Given its focus on the core-experience of light, this book is as you have no doubt noticed a strange history. It does not move so much across time, propelled by cause and effect, as into it, motivated by experience and its imaginative interpretation. *Light as Experience and Imagination* has simply described various episodes involving past encounters with light. We have probed patterns of the mind generated by these, revealed in their resulting artifacts of light, from megalith to treatise. On occasion we have observed influences of one person's experience of light on another, such as Parmenides' coming to terms with light-shadow appearances on Plato, or on an entire later culture. But because light-experience does not improve or get closer to reality, we have not in the main traced any path towards a single truth. Rather, the story of light as experience insists on being episodic because all human relationships with light are equally valid. This book is thus a modest enterprise, having merely examined successively new aspects of light's character, no grand narrative. So I cannot conclude here that we have arrived at some ultimate truth but only found ourselves in possession of a richly textured fabric of truths. No one experience gets closer to the "real essence" of light or its "secret nature," which histories of advances in the study of the nature of light and vision (optical science) take as their mission.

In fact, this book denies the assumption underlying such a mission, that light "out there," as if a stone, is distinct from light "in here," its representation in the mind. Rather, our light-experiences occupy a space between these, both being rooted in our individual physiology and psychology. Thus we lack the ability in our relationships with light to distinguish between the "real light" of science (optical/physiological) and the "phenomenal light" of experience (perceptual/conceptual), because the body is the locus of all encounters with light. After all, "we eat light," as the light-space artist James Turrell reminds us, "drink it in through our skins."[3] In truth, in the world of light as experience and imagination there is no distinction between real and phenomenal. Light is one.

Proceeding episodically, however, does not mean our encounters over time with light in the space between "in here" and "out there" have no continuity. As cultures change, of course, so do light-experiences. Yet what does not change is our susceptibility to light. Akhenaten felt this to such a degree that for him light seemed to rejoice in its own beauty as if possessing a "self-awareness." In all ages light has drawn us into intimacy, sometimes against our will as it did Augustine, as if we recognized with him on a bodily level our own mind coming into an "awareness of itself."[4]

Yet the lighteaters we've met throughout history have at the same time felt light to be a self-contained substance "out there," independent of source, such as the sun, even when known, and of creation itself. Hence we have been compelled to steal brightness, as did the shaman, or receive rays, as did the seer. Light was for them as well as their successors an independent and sensual character in the world. We have been moved to welcome naked "splendid" light, as did Akhenaten, a self-created light that the Luminaries and Refractors gave full expression to. And we have felt compelled to enter into a dialogue with light, as did Aeschylus' with light-shadow. And like the mosaicist Dioscurides five centuries later induced to become intimate with a light acting on its own. This was for him a light acting on light; for Lucretius it was a flash striking without, and for Plotinus within. The "torrent of rays" from the noon sun that filled Philo's "whole soul," we recall, acted independently of his will.

We have seen continuity as well in persistent manifestations of light revisited but given new significance in successive ages. Salient facets of light's character such as brightness have universally intruded into our experience, although it has been invested with different values in each age, from the shamans who experienced it as sheer power to the luminaries who resonated with its inexhaustible activity. Similarly, light's relationship with darkness has taken on various values in experience. In prehistoric eras, light and darkness were complementary, a relationship revalued by priest-kings who charged it with moral significance—good and evil—which ultimately hardened into a dualism. This was only to be revalued, as we've seen, in the Classical period. Luminaries from Thales to Aeschylus participate in the colloquy of light and shadow. Neither complementary nor adversarial, light and darkness in their experience enter a contrary relationship. One does not negate the other; each defines the other. They exist in a relationship of mutual questioning, quietly with the vase-painters, loudly with the Dionysiac drama of Aeschylus. Yet this is revalued once again by the Refractors who experience light as a free agent. If the Greeks were preoccupied with light in relation to shadow, as I've said, the Romans were obsessed with light in relation to themselves. Hence we've seen that their respective interpretations of core-experience reveal something of light's psychology, the relationship of mind and world, as well as its biography, the relation between an individual's life-story and light's history.

While certain manifestations of light thread our story from the beginning, we've seen at the same time discontinuities from age to age. Although the common and simple, glorious and mysterious character of light has always been evident to our witnesses, as it was to Reverend Hunter, we have seen profound shifts in the consciousness of light. Prehistoric peoples' experience was largely subconscious, dealing with light as a common and simple phenomenon in much the same way they had to deal with the flora and fauna of their lands. But for the cosmopolitan peoples of Egypt and neighboring countries, the emphasis shifted to an overtly conscious one. With this came the flowering of an awareness of the aesthetic and spiritual dimensions of light's character, its gloriousness and mysteriousness. And light rejoices in its sacred power, which was exploited by the Egyptians. Light now revels in its own beauty, a trait fully celebrated in classical times by the Greeks and Romans. This shift in awareness from prehistoric eras to Classical ages parallels that from the preoccupation with brightness and rays, the sensual and independent character of light, to a near obsession with radiance and energy, the self-aware and active character. And this coincided, you may recall, with the great historical threshold crossed in our cultural evolution from mythic to theoretic consciousness.

Within these major movements from age to age we also discovered notable shifts in the interpretation of light-experience. Finding different expressions, respectively, in paint and stone, the shaman's encounter with light was a visionary one, whereas the seer's was one of vision. The former via altered states of consciousness, *being* light; the latter via ordinary consciousness, *seeing* light. And we heard the testimony of a Pharaoh's relation to light that was both visionary and visioning, a two-way gaze, as Akhenaten restores the initial physical encounter with natural light, its sheer presence and beauty. Later the Greeks of the Classical period were compelled by this presence to explore its "psychology," exposing light's optical character braided with the senses as well as its trickster role insinuated with the mind. Probing light's make-up, the Greeks, from Thales to the red- and black-figure vase artists to the theatre painters and Euclid, seized mainly on the predicable aspects of light's character. And this discovery led to the later painters Agatharcus and Apollodorus' desire to portray light itself for the first time in history, introducing a "depicted light" that the Roman artists further explored.

Arising from a shift in the experience of light, the Romans were bent on discovering its imaginative "biography," in which they grounded their relationship. For Plotinus light is aware of itself, "light sees light." He encounters a free, self-conscious character in the world. The mosaic and wall artists discovering its shiftiness and Lucretius its randomness, then, found expression for its unpredictable aspects. This free agent demanded portrayal as such. And the mosaic and wall painters obliged, giving us a self-conscious character in their portrait of a "depicted" light's relationship with the "actual" light, which has shaped the Western experience and imagination of light to this day. The unpredictable behavior that came with light's agency was countered by the Romans who probed its history, ultimately making light's foundational origins their own. If the Greeks observed a static light from a distance, as if in a theatre, preoccupied with its predictable behavior, that is, which Aeschylus challenged (Dionysus) and confirmed (scene-painters) at the same time, the Romans coveted intimacy. And thus their painters gave light a portrait, and their writers traced its biography. They were absorbed by its subjective aspects, the convicting power of Hadrian's heliophany, and the inner dynamic of Gnostic light and John the Evangelist' logos-light.

WHAT WE KNOW OF LIGHT'S CHARACTER AND WHY IT MATTERS

At this point it is evident that any generalizations about light's character over time are undercut by its human involvement with ourselves as individual knowers within time and language, being a function of personal biographies and exigent conditions of geography and culture. Lacking a unitary conclusion about our relationship with light and its morphing over time, then, we turn to a couple of basic questions that yet may haunt us: What do we really know about light? Why does it matter? Approaching those lighteaters who have shed light on light in our investigation, we can return to Reverend Hunter's summary of light's character as a framework for some answers.

Light being radically democratic in its commonness, filling every nook and cranny of our world and appearing to everyone without distinction, we have seen that what we know is bound up with what we hear in personal testimony—interpretations—from individuals, each shedding light on light. And what we hear in their testimonies, whether in paint or prose, is light telling us about itself, revealing aspects of its character. This is all we have because, of course, we cannot touch light, although light is nothing less than the world, objects themselves being in a real sense "worn-out light." They reveal themselves to us, that is, only because they absorb light or bounce it to us. We remember the Shaman's desire to intercept light before it is worn out, to experience it as a being that he merges with. And Lucretius' encounter with its flash and glint, which in contrast he keeps at a distance. All either could do is take light at its word in their lives, the triggering image, and construct patterns from this core-experience to explain it to themselves. They realize that they must start from the stark realization that "it alone says without saying."[5]

And what light says is simplicity itself. It reveals a self-contained character who acts independently, every one of our witnesses testifies, energetically radiating from all points (whether directly from a source or indirectly by reflection or diffusion). At the same time, entwined with our lives, light encompasses a complex range of human interpretation from the "spiritual" of the shaman's to the "factual" of the Greek's, such as Thales'. Accordingly, this character cuts across fact, erasing the line between the experience of those who come to believe in it (religion) and those who want to explain it (science), between a passive and active encounter with light. The experience of each of our witnesses is, as we've seen, necessarily both. That is, Light's very character challenges us to abandon cherished notions of fundamental divides in the world. And we are thrust back on our imaginative interpretation—patterns in the mind—because in the end, as we've learned, light consists fundamentally of propositions about light.

In the core-experiences of light we've seen, there is no divide between admiring and studying light, between the physical and the spiritual, literal and figurative, as shamans and Luminaries alike found, which was unsettling for Plato and Augustine. It erases spatial (inside-outside) and temporal (contingent-deterministic) boundaries, respectively, for the wall painters and Lucretius. Likewise, light becoming the personage of man-God (Jesus/Logos) erases spiritual borders for John the Evangelist, and being "pure presence" erases corporeal-incorporeal ones for Plotinus. For the image triggers an imaginative response, which exploits metaphors educing both sides. This dynamic accounts for testimonies of a protean light having many faces, even being contradictory. Yet its being a democratic and constant character inseparable from our experiencing self is what in the end we know about light.

But if light is merely propositions about light, what have we gained by attending to individual testimonies of light-experiences down through the ages? As you've discovered in this history, we experience light physically, emotionally, and intellectually as self-generated, evoking the very image it gives birth to in our experience. And because this image becomes a "site of possibilities and departures" for imaginative interpretation that is at once its very drive and arrival, light reveals us to ourselves.[6] By our relationship with this paradoxically simple and complex character, giving it an air of mystery that many interpret as sacred, light reveals our authentic self. One co-evolved with light as it generates images in its interaction with matter, being reflected by a river or diffused by the sky. Our eyes are products of light just as light is a product of our experience.

My history of encounters with light, then, has exposed thread by interpretive thread, woven over-under and into cords of core light-experiences, fundamental dimensions of human consciousness. Accumulated stories of the relationships with light we've heard have revealed layers of light experience-imagination from Paleolithic to Roman times. These do not progress, approaching a final truth, as we've learned, but rather yield a richly layered tapestry of human experience. This expands our consciousness of what it is to be human in our relation to the world. Namely, the complexity of our interaction with light defines us to a large degree as human beings. This also gives us an enlarged sense of our imaginative powers, the realization that our needs and desires are matched by an unmatched interpretive imagination.

And finally, this layering of light-experience deepens our knowledge and understanding of the power of beauty. It is primarily light's beauty that seduces us, some falling for it completely as did Plotinus, or resisting it as did Augustine. But after hearing testimonies from these and other sages of the Classical Age, as well as those of shamans, seers, and kings, I suspect we would agree that the world reveals more beauty the more that light is known to us. My hope is that this is reward enough, if not the single conclusion we may desire, for time spent with this book.

NOTES

1. "Profane universe," Mircea Eliade, "Spirit, Light, and Seed" (*History of Religions* 11.1 [Aug., 1971]), 2; "preferred key," Matthew R. Kapstein, "Rethinking Religious Experience" (*The Presence of Light: Divine Radiance and Religious Experience* [Chicago: University of Chicago Press, 2004]), 280.
2. Jacques Roubaud, *Exchanges on Light* (Iowa City: La Presse, 2009), 26.
3. "Real light," "phenomenal light," Maurice Merleau-Ponty quoted by Cathryn Vasseleu, *Textures of Light: Vision and Touch in Irigaray, Levinas and Merleau-Ponty* (London: Routledge, 1998), 42–43; "we eat light," James Turrell, quoted by Michael Govan, "Inner Light: The Radical Reality of James Turrell," *James Turrell: A Retrospective*, Los Angeles County Museum of Art, 13. Accessed April 31, 2014. https://www.randomhouse.de/leseprobe/James-Turrell-A-Retrospective/leseprobe_9783791352633.pdf.
4. Kapstein, "Rethinking Religious Experience," 285.
5. Roubaud, *Exchanges on Light*, 9, 29.
6. Roubaud, *Exchanges on Light*, 47.

Bibliography

INTRODUCTION, GENERAL SCOPE AND INTEREST

This selection includes books and articles referenced in the notes as well as a number providing essential background. Others have influenced my approach in general and illuminated the "languages of light." Items marked with an asterisk (*) are of central importance as well as of interest for further reading.

Acland, James. "Light as Architecture." *Artscanada* XXV.5 (Issue 124/127, 1968): 54–58.

Arnheim, Rudolf. *Art and Visual Perception*. Berkeley: University of California Press, 1971.

Atkins, Peter W. *Creation Revisited*. New York: W. H. Freeman, 1992.

Bachelard, Gaston. "The Painter Solicited by the Elements." In *The Right to Dream*. Translated by J. A. Underwood. New York: Grossman, 1971.

Bainbridge, Richard, et al., eds. *Light as an Ecological Factor*. London: Blackwell, 1966.

Bal, Mieke. "Light in Painting: Dis–Seminating Art History." In *Deconstruction and the Visual Arts: Art, Media, Architecture*, ed. Peter Brunette and David Wills. Cambridge: Cambridge University Press, 1994.

Bell, Janis Callen. "Light: Pictorial Representation of Light." In *The Dictionary of Art (Grove)*, vol. 19, ed. Jane Turner. New York: MacMillan, 1996.

Bille, Mikkel, and Tim Flohr Sørensen. "An Anthropology of Luminosity: The Agency of Light." *Journal of Material Culture* 12.3 (2007): 263–84.

Birren, Faber. "The Significance of Light." *American Institute of Architects Journal* 58.2 (August 1972): 16–19.

Bixby, F. "A Phenomenological Study of Luster." *Journal of General Psychology* 1 (1928): 136–74.

Blassnigg, Martha, ed. *Light Image Imagination*. Amsterdam: Amsterdam University Press, 2013.

*Blumenberg, Hans. "Light as a Metaphor for Truth." In *Modernity and the Hegemony of Vision*, ed. David Michael Levin. Berkeley, CA: University of California Press, 1993.

Bova, Ben. *The Beauty of Light*. New York: John Wiley & Sons, 1988.

Boyer, Carl. B. *The Rainbow: From Myth to Mathematics*. New York: T. Yoseloff, 1959.

Branch, Mark Alden. "Light Architecture." *Progressive Architecture* 73.11 (November 1992): 76–79.

Brill, Thomas B. *Light: its Interaction with Art and Antiquities*. New York: Plenum Press, 1980.
Bronowski, Jacob. *The Common Sense of Science*. Cambridge: Harvard University Press, [1953] 1978.
Brox, Jane. *Brilliant: The Evolution of Artificial Light*. Boston: Houghton Mifflin Harcourt, 2010.
Churchland, Paul M. "The Rediscovery of Light." *Journal of Philosophy* 93.5 (May 1996): 211–28.
Clark, Kenneth. *Landscape Into Art*. Boston: Beacon Press, 1961.
Clegg, Brian. *Light Years: An Exploration of Mankind's Enduring Fascination with Light*. London: Piatkus, 2001.
Collis, John Stewart. *The World of Light*. New York: Horizon Press, 1960.
Dictionary of National Biography. Edited by Lord Blake and C. S. Nicholls. New York: Oxford University Press, 1986.
Dictionary of Untranslatables: A Philosophical Lexicon. Edited by Barbara Cassin. Translated by Steven Rendall, et al. Princeton, NJ: Princeton University Press, 2014.
Ditchburn, R. W. *Light*. Third ed. New York: Academic Press, 1976.
Dogigli, Johannes. *The Magic of Rays*. Edited and translated by Charles Fullman. New York: Knopf, 1961.
Edwards, Carolyn McVickar. *The Return of the Light: Twelve Tales from Around the World for the Winter Solstice*. New York: Marlowe, 2000.
Eliade, Mircea. "Spirit, Light, and Seed." *History of Religions* 11.1 (August 1971): 1–30.
Encyclopedia of Comparative Iconography. Edited by Helene E. Roberts. Chicago: Fitzroy Dearborn, 1998.
Evans, Benjamin H. *Daylight in Architecture*. New York: McGraw Hill, 1981.
Feinberg, Gerald. "Light." *Scientific American* 219.3 (September 1968): 51.
Flam, Jack. *Matisse on Art*. Rev. ed. Berkeley: University of California Press, 1995.
Foucault, Michel. *The Archaeology of Knowledge*. Translated by A. M. Sheridan Smith. New York: Pantheon, 1972
Frölich, Fanchon. "The Locations of Light in Art: from Rembrandt to Op Art and Light Environment." *British Journal of Aesthetics* 11.1 (Winter 1971): 48–62.
Frye, Northrop. "The Light of Myth." *Artscanada* 25.5, Issue 124/127 (1968): 8.
Gibson, James J. *The Senses Considered as Perceptual Systems*. Boston: Houghton–Mifflin, 1966.
Gifford, Don. *A Natural History of Perception: The Farther Shore*. New York: Vintage, 1990.
Gombrich, E. H. *Art and Illusion*. Second ed. New York: Pantheon, 1961.
*Grandy, David. "The Otherness of Light: Einstein and Levinas." *Postmodern Culture* 12.1 (September 2001): sections 1–47. Accessed August 17, 2011. http://pmc.iath.virginia.edu/issue.901/12.1grandy.html.
Gross, Michael. *Light & Life*. New York: Oxford University Press, 2002.
Harris, Alexandra. *Weatherland: Writers and Artists Under English Skies*. London: Thames & Hudson, 2015.
van Heel, A. C. S., and C. H. F. Velzel. *What is Light?* Translated by J. L. J. Rosenfeld. London: Weidenfeld and Nicolson, 1968.
Hendrick, Ellwood. "Mystery and Power of Light." *Atlantic Monthly*, 133 (May 1924): 618–24.
Henri, Robert. *The Art Spirit*. Boulder, CO: Westview Press, 1923.
Hess, Thomas B., and John Ashbery, eds. *Light: From Aten to Laser*. *Art News Annual, XXXV*. New York: Macmillan, 1969.
Home, Dipankar, and John Bribbin. "What is Light?" *New Scientist* 132.1793 (November 2, 1991): 30–33.
Hunter, Henry. Sermon VI (1793). In *Sermons Preached at Different Places and on Various Occasions*, vol. 2. London: Printed for the Author, 1795.
Ings, Simon. *A Natural History of Seeing: The Art and Science of Vision*. New York: W.S. Norton, 2008.
Johnson, Charles. *The Language of Painting*. New York: Greenwood Press, 1969 [1949].

Jütte, Robert. *A History of the Senses from Antiquity to Cyberspace*. Translated by James Lynn. Malden, MA: Polity Press, 2005.

*Kapstein, Matthew R., ed. *The Presence of Light: Divine Radiance and Religious Experience*. Chicago: University of Chicago Press, 2004.

Kaufman, Donald, and Taffy Dahl. *Color and Light*. New York: Clarkson Potter, 1999.

Keller, Max. *Light Fantastic: The Art and Design of Stage Lighting*. New York: Prestel, 1999.

Kemp, Martin. *The Science of Art: Optical Themes in Western Art from Brunelleschi to Seurat*. New Haven, CT: Yale University Press, 1990.

Kepes, Gyorgy. *Light as a Creative Medium*. Cambridge: Harvard University Press, 1965.

Lamb, Trevor D. "Evolution of the Eye." *Scientific American* 305.1 (July 2011): 64–69.

"Light." *Artscanada* 124/127 (December 1968).

"Light." In *Man, Myth & Magic: An Illustrated Encyclopedia of the Supernatural*, vol. 12, ed. Richard Cavendish. New York: Marshall Cavendish, 1970.

"Light." *Scientific American* 219.3 (September 1968).

"Light." *Semiotica* 136.1–4 (2001).

Livingston, Margaret S. *Vision and Art: The Biology of Seeing*. New York: Harry M. Abrams, 2002.

Lynch, David K., and William Livingston. *Color and Light in Nature*. Second ed. London: Cambridge University Press, 2001.

Major, Mark, Jonathan Speirs, and Anthony Tischhauser. *Made of Light: The Art of Light and Architecture*. Basel, Switzerland: Birkhäuser, 2005.

Mason, Peter. *Light Fantastic*. London: Penguin, 1981.

Merleau–Ponty, Maurice. *The Visible and the Invisible*. Translated by Alphonso Lingis. Edited by Claude Lefort. Evanston, IL: Northwestern University Press, 1968.

Meyers, Victoria. *Designing with Light*. London: Laurence King, 2006.

Minnaert, M. *The Nature of Light & Colour in the Open Air*. New York: Dover, 1954.

Morris, Richard. *Light*. Indianopolis: Bobbs–Merrill, 1979.

Mueller, Conrad G., and Mae Rudolph. *Light and Vision*. New York: Time Inc., 1966.

*Park, David. *The Fire within the Eye*. Princeton, NJ: Princeton University Press, 1997.

Perkowitz, Sidney. *Empire of Light: A History of Discovery in Science and Art*. New York: Henry Holt, 1996.

Plummer, Henry. *The Architecture of Natural Light*. New York: Random House/Monacelli Press, 2009.

Plummer, Henry. *Masters of Light: Twentieth-Century Pioneers*. Tokyo: a+u Extra Edition, 2003.

*Plummer, Henry. *Poetics of Light*. (*Architecture and Urbanism, 1987*, December Extra Edition) Tokyo: a+u Publishing, 1987.

Ponzio, Julio. "Light, Being, and Time." *Semiotica* 136.1–4 (2001), 55–67.

Roberts, Helene E. "Light II: Divine, Natural, and Neon." In *Encyclopedia of Contemporary Iconography*. Chicago: Fitzroy Dearborn, 1998.

Ronchi, Vasco. *The Nature of Light: An Historical Survey*. Translated by V. Barocas. Cambridge, MA: Harvard University Press, 1970.

*Roubaud, Jacques. *Exchanges on Light*. Translated by Eleni Sikelianos. Iowa City, IA: La Presse, 2009.

Rublowsky, John. *Light: Our Bridge to the Stars*. New York: Basic Books, 1964.

Ruechardt, Eduard. *Light: Visible and Invisible*. Ann Arbor, MI: University of Michigan Press, 1958.

Russell, Sage. *The Architecture of Light*. La Jolla, CA: Conceptnine, 2008.

Schwartz, Martin. "Light Organizing / Organizing Light." *Places* 8.2 (1992): 14–25.

Serres, Michel. *A History of Scientific Thought. Elements of a History of Science*. Cambridge, MA: Blackwell, 1995.

Shlain, Leonard. *Art and Physics: Parallel Visions in Space, Time and Light*. New York: William Morrow and Co., 1991.

Silverman, Mark P. *Waves and Grains: Reflections of Light and Learning*. Princeton, NJ: Princeton University Press, 1998.

Singh, Madanjeet. *The Sun: Symbol of Power and Life*. New York: H. N. Abrams, 1993.

*Sobel, Michael I. *Light*. Chicago: University of Chicago Press, 1987.
Stevens, Wallace. *The Collected Poems*. New York: Knopf, 1972.
Summers, David. *Real Spaces*. New York: Phaidon, 2003.
———. "Real Metaphor: Towards a Redefinition of the 'Conceptual Image.'" In *Visual Theory: Painting and Interpretation*, ed. Norman Bryson, et al. New York: HarperCollins, 1991.
Tanizaki, Junichirō. *In Praise of Shadows*. Translated by Thomas J. Harper and Edward G. Seidensticker. Tokyo: Tuttle, 1977.
Tolansky, Samuel. *Curiosities of Light Rays and Light Waves*. New York: American Elsevier, 1965.
Turner, Jane. *The Dictionary of Art*. 34 vols. New York: Grove, 1996.
Tymieniecka, Anna-Teresa, ed. *The Elemental Dialectic of Light and Darkness: The Passion of the Soul in the Onto–Poiesis of Life*. Boston: Kluwer Academic, 1992.
Vasseleu, Cathryn. *Textures of Light: Vision and Touch in Irigaray, Levinas and Merleau–Ponty*. London: Routledge, 1998.
Vavilov, Sergei I. *The Human Eye and the Sun*. Translated by O. M. Blum. Edited by S. Tolansky. Oxford: Pergamon, 1965.
*Watson, Bruce. *Light: A Radiant History from Creation to the Quantum Age*. New York: Bloomsbury, 2016.
Wegscheider, Jane Hyman. *The Light Book: How Natural and Artificial Light Affect Our Health, Mood, and Behavior*. Los Angeles: Jeremy P. Tarcher, 1990.
Weightman, Barbara A. "Sacred Landscapes and the Phenomenon of Light." *The Geographical Review* 86.1 (January 1996): 59–71.
Weinberg, Steven. *The First Three Minutes*. New York: Basic Books, 1977.
Weiss, Richard. *A Brief History of Light and Those That Lit the Way*. Singapore: World Scientific Publishing.
*Zajonc, Arthur. *Catching the Light: The Entwined History of Light and Mind*. New York: Bantam, 1993.
Zucker, Wolfgang M. "In the Light of Being." *Journal of Aesthetics and Art Criticism* 27.2 (Winter 1968): 156–57.

CHAPTER 1: PALEOLITHIC ERA

Aujoulat, Norbert. *Lascaux: Movement, Space, and Time*. New York: Abrams, 2005.
Bahn, Paul G., and Jean Vertut. *Journey Through the Ice Age*. Berkeley: University of California Press, 1997.
Bataille, George. *Lascaux or the Birth of Art*. Cleveland, OH: World Publishing, n.d.
Beltrán, Antonio, ed. *The Cave of Altamira*. Photographs by Pedro A. Saura Ramos. New York: Harry M. Abrams, 1998.
Bernaldo de Quirós, Federico. "The Cave of Altamira: Its Art, Its Artists, and Its Times." *The Cave of Altamira*, ed. Antonio Beltrán. New York: Harry M. Abrams, 1998.
Boas, Franz, and Henry W. Tate. "Tsimshian Mythology." In *Thirty–First Annual Report of the Bureau of American Ethnology, 1909–1910*. Washington: Government Printing Office, 1916.
Breuil, Henri. *Four Hundred Centuries of Cave Art*. Montignac: Centre d'études et de Documentation Préhistoriques, 1952.
Brox, Jane. *Brilliant: The Evolution of Artificial Light*. New York: Houghton Mifflin, 2010.
Campbell, Joseph. *The Masks of God: Primitive Mythology*. New York: Viking, 1970.
Chauvet, Jean-Marie. *Dawn of Art: The Chauvet Cave*. New York: Abrams, 1996.
Clausen, Constance. "Creation by Sound / Creation by Light: A Sensory Analysis of Two South American Cosmologies." In *The Varieties of Sensory Experience: A Sourcebook, ed.* David Howes. Toronto: University of Toronto Press, 1991.
Clottes, Jean. "Art of the Light and Art of the Depths." In *Beyond Art: Pleistocene Image and Symbol*, ed. Margaret W. Conkey, et al. San Francisco: The California Academy of Sciences, 1997.

———. *Chauvet Cave: The Art of Earliest Times*. Translated by Paul G. Bahn. Salt Lake City: University of Utah Press, 2003.

———. *What is Paleolithic Art? Cave Paintings and the Dawn of Human Creativity*. Translated by Oliver Y. Martin and Robert D. Martin. Chicago: University of Chicago Press, 2016.

Codrington, Robert Henry. "Religious Beliefs and Practices in Melanesia." *The Journal of the Anthropological Institute of Great Britain and Ireland* 10 (1881): 271–72.

Conkey, Margaret W. *Beyond Art: Pleistocene Image and Symbol*. Berkeley: University of California Press, 1997.

———. "Making Things Meaningful: Approaches to the Interpretation of the Ice Age Imagery of Europe." *Meaning in the Visual Arts*, ed. Irving Lavin. Princeton, NJ: Institute for Advanced Study, 1995, 49–64.

Corless, Roger. "After Eliade, What?" Review of Mircea Eliade, *Essential Sacred Writings from Around the World* in *Religion* 23.4 (1993): 373–77.

Cruikshank, William. "Light and Darkness." *Hastings Encyclopedia of Religion and Ethics*. 13 vols., ed. James Hastings and J. A. Selbie. New York: Scribner's, 1901–1926.

Curtis, Gregory. *The Cave Painters: Probing the Mysteries of the World's First Artists*. New York: Knopf, 2006.

Doria, Charles, and Harris Lenowitz, eds. *Origins: Creation Texts from the Ancient Mediterranean*. Garden City, NY: Doubleday, 1976.

Eliade, Mircea. *Cosmos and History*. New York: Harper & Row, 1959.

———. *Patterns in Comparative Religion*. Translated by Rosemary Sheed. New York: New American Library, 1958.

———. *The Two and the One*. Translated by J. M. Cohen. New York: Harper, 1965.

Ellwood, Robert E. *The Politics of Myth: A Study of C.G. Jung, Mircea Eliade, and Joseph Campbell*. Albany, NY: State University of New York Press, 1999.

Encyclopedia of Creation Myths. Edited by David Adams Leeming with Margaret Adams Leeming. Santa Barbara, CA: ABC-CUO, 1994.

Fagan, Brian. *Cro-Magnon: How the Ice Age Gave Birth to the First Modern Humans*. New York: Bloomsbury Press, 2010.

Freeman, L. G., et al., eds. *Altamira Revisited and Other Essays on Early Art*. Chicago: Institute for Prehistoric Investigations, 1987.

Giedion, S. *The Eternal Present: The Beginnings of Architecture*. Mellon Lectures XXXV, Part 2. New York: Pantheon. 1964.

Grand, P. M. *Prehistoric Art*. Greenwich, CT: New York Graphic society, 1967.

Handbook of North American Indians: Vol 7, Northwest Coast, ed. Wayne Suttles. Washington: Smithsonian Institution, 1990.

Hawkes, Jacquetta. *Man and the Sun*. New York: Random House, 1962.

Hayden, Brian, and Suzanne Villeneuve. "Review, *Lascaux* by Norbert Aujoulat and *Chauvet Cave* by Jean Clottes." *American Antiquity* 70.2 (2005): 384–88.

Howes, David, ed. *The Varieties of Sensory Experience: A Sourcebook*. Toronto: University of Toronto Press, 1991.

Hyde, Lewis. *Trickster Makes This World: Mischief, Myth, and Art*. New York: Farrar, Straus and Giroux, 1998.

Jensen, Allan. "A Structural Approach to the Tsimshian Raven Myths: Lévi–Strauss on the Beach." *Anthropologica*, n.s., 22.2 (1980): 159–86.

Jolly, Pieter. "On the Definition of Shamanism." *Current Anthropology* 46.1 (February 2005): 127–28.

Kanowski, M. G. "Towards an Understanding of the X–Ray Technique in Art: Clues from Other Art Traditions." *Oceania* XLIV.2 (December 1973): 96–100.

Larsen, Stephen. *The Shaman's Doorway*. New York: Harper & Row, 1976.

Lavin, Irving, ed. *Meaning in the Visual Arts: Views from the Outside*. Princeton, NJ: Institute for Advanced Study, 1995.

Layton, Robert. "Review: *The Mind in the Cave: Consciousness and the Origins of Art*." *Antiquity* 77.296 (June 2003): 422.

Leach, Maria. *The Beginning*. New York: Funk and Wagnalls, 1956.

Leeming, David Adams, and Margaret Adams Leeming. *Encyclopedia of Creation Myths*. Santa Barbara, CA: ABC-CLIO, 1994.

Leroi–Gourhan, André. *Treasures of Prehistoric Art*. Translated by Norbert Guterman. New York: Abrams, 1967.

Lewis-Williams, David. *The Mind in the Cave: Consciousness and the Origins of Art*. London: Thames & Hudson, 2002.

———. *Inside the Neolithic Mind: Consciousness, Cosmos and the Realm of the Gods*. London: Thames & Hudson, 2005.

———, et al. "Threads of Light: Re–Examining a Motif in Southern African San Rock Art." *The South African Archaeological Bulletin* 20.1 (June 1988): 123–36.

Lommel, Andreas. *Prehistoric and Primitive Man*. New York: McGraw-Hill, 1966.

Marriott, Alice, and C. K. Rachlin, eds. *American Indian Mythology*. New York: Thomas Y. Crowell, 1968.

McDonald, Maggie. "Pieces of Light." Interview with Nick Saunders. *New Scientist* 168.2262 (October 2000): 42–45.

Michaux, H. *Light Through Darkness*. New York: Orion, 1963.

Mithen, Steven J. *After the Ice: A Global Human History, 20,000-5000 B.C.* Cambridge: Harvard University Press, 2004.

———. "A Creative Explosion? Theory of Mind, Language and the Disembodied Mind of the Upper Paleolithic." In *Creativity in Human Evolution and Prehistory*, ed. Steven J. Mithen. London: Routledge, 1998.

Nechvatal, Joseph. "Immersive Excess in the Apse at Lascaux." *Technoetic Arts* 3.3 (2005): 181–92.

———. "Jean Baudrillard and a Counter–Mannerist Art of Latent Excess." Accessed November 30, 2011. http://www.eyewithwings.net/nechvatal/Baudrillard/baudrillard.htm.

Nowell, April. "From a Paleolithic Art to Pleistocene Visual Cultures." *Journal of Archaeological Method and Theory* 13.4 (December 2006): 239–49.

Olcott, William Tyler. *Sun Lore of All Ages*. New York: Dover, 2005 [1914].

Oxford Companion to World Mythology, ed. David Leeming. New York: Oxford University Press, 2005.

Pérez-Seoane, Matilde Múzquiz. "Technique, Individual Artists, and Artistic Concepts in the Painting of Altamira." In *The Cave of Altamira* edited by Antonio Beltrán. New York: Harry M. Abrams, 1998.

Pfeiffer, John E. *The Creative Explosion: An Inquiry into the Origins of Art and Religion*. New York: Harper & Row, 1982.

Powell, T. G. E. *Prehistoric Art*. New York: Praeger, 1966.

Raphael, Max. *Prehistoric Cave Painting*. Translated by Norbert Guterman. The Bollingen Series IV. New York: Pantheon, 1945.

Reid, Bill, and Robert Bringhurst. *The Raven Steals the Light*. Seattle, WA: University of Washington Press, 1996.

Robert-Lamblin, Joëlle. "An Anthropological View." In *Chauvet Cave: The Art of Earliest Times*, ed. Jean Clottes. Translated by Paul G. Bahn. Salt Lake City: University of Utah Press, 2003.

Robinson, John. "The Cave of Chauvet-Pont-d'Arc." Accessed June 23, 2011. http://www.culture.gouv.fr/culture/arcnat/chauvet/en/index.html.

Ruspoli, Mario. *The Cave of Lascaux: The Final Photographs*. New York: Abrams, 1987.

Ryba, Thomas. "Elemental Forms, Creativity and the Transformative Power of Literature in A-T. Tymieniecka's *Tractatus Brevis*." In *The Elemental Dialectic of Light and Darkness: The Passion of the Soul in the Onto-Poiesis of Life*, ed. *A-T Tymieniecka*. Norwell, MA: Kluwer Academic, 1992.

Saliba, John A. *Homo Religiosus in Mircea Eliade: An Anthropological Evaluation*. Leiden: E. J. Brill, 1976.

Sandars, N. K. *Prehistoric Art in Europe*. Baltimore, MD: Penguin, 1968.

Saunders, Nicholas J. "Stealers of Light, Traders in Brilliance: Amerindian Metaphysics in the Mirror of Conquest." *Res: Anthropology and Aesthetics* 33 (1998): 225–52.

Scott, Katie. "Neanderthals Mated with Humans Outside of Africa," *Wired*, July 18, 2011. Accessed January 24, 2012. http://www.wired.com/wiredscience/2011/07/neanderthal–human–mating/#more–67842.

Sieveking, Ann. *The Cave Artists*. London: Thames and Hudson, 1979.

Smith, Noel. *An Analysis of Ice Age Art: Its Psychology and Belief System*. New York: Peter Lang, 1992.

Sproul, Barbara C. *Primal Myths; Creating the World*. New York: Harper & Row, 1979.

Standard Dictionary of Folklore, Mythology, and Legend. Vol. II, "Light." New York: Funk & Wagnalls, 1950.

Thompson, Stith. *Tales of the North American Indians*. New York: Midland, 1957 [1929].

———, ed. *Motif-Index of Folk-Literature*. Rev. and enl. Bloomington: Indiana University Press, 1966.

Torbrügge, Walter. *Prehistoric European Art*. New York: Abrams, 1968.

Tymieniecka, Anna-Teresa, ed. *The Elemental Dialectic of Light and Darkness: The Passion of the Soul in the Onto-Poiesis of Life*. Norwell, MA: Kluwer Academic, 1992.

Ucko, Peter J., and Adrée Rosenfeld. *Palaeolithic Cave Art*. New York: McGraw-Hill, 1967.

Van Over, Raymond. *Sun Songs: Creation Myths*. New York: Plume, 1980.

Vasseleu, Cathryn. *Textures of Light: Vision and Touch in Irigaray, Levinas and Merleau–Ponty*. London: Routledge, 1998.

Whitley, David. "The Cave of Chauvet-Pont-d'Arc." Accessed June 23, 2011. http://www.culture.gouv.fr/culture/arcnat/chauvet/en/index.html.

———. "Witnesses." Accessed November 30, 2011. http://www.culture.gouv.fr/culture/arcnat/chauvet/en/index.html.

Wiebe, Donald. "Roots in the Brain: On David Lewis-Williams and David Pearce, *Inside the Neolithic Mind*." In *Contemporary Theories of Religion*, ed. Michael Stausberg. New York: Routledge, 2009.

CHAPTER 2: NEOLITHIC ERA

Armstrong, Karen. *A Short History of Myth*. Edinburgh: Canongate, 2005.

Aveni, Anthony F., ed. *Archaeoastronomy in Pre-Columbian America*. Austin: University of Texas, Press, 1975

———. *Stairway to the Stars: Skywatching in Three Great Ancient Cultures*. New York: Wiley, 1997.

Bachelard, Gaston. *The Poetics of Space*. Translated by Marie Jolas. Boston: Beacon, 1969.

Bednarik, Robert. "Cognition and Symbolism in Human Evolution [Lecture No. 3, Semiotix Course 2006]." Accessed August 25, 2011. http://projects.chass.utoronto.ca/semiotics/cyber/rbednarik3.pdf.

Bellah, Robert N. *Religion in Human Evolution: From the Paleolithic to the Axial Age*. Cambridge: Harvard University Press, 2011.

Bradley, Richard J. "Architecture, Imagination and the Neolithic World." *Creativity in Human Evolution and Prehistory*, edited by Steven Mithen. London: Routledge, 1998.

———. "Deaths and Entrances: A Contextual Analysis of Megalithic Art." *Current Anthropology* 30 (1989): 68–75.

———. "Ritual, Time and History." *World Archaeology* 23.2, Chronologies (October 1991): 209–19.

Budge, E. A. W. *Babylonian Life and History*. London: The Religious Tract Society, 1884.

Burl, Anthony. *The Stone Circles of the British Isles*. New Haven, CT: Yale University Press, 1976.

Campbell, Joseph. *The Masks of God: Primitive Mythology*. New York: Viking, 1970.

———. *The Mythic Image*. Princeton, NJ: Princeton University Press, 1974.

Caspari, Rachel. "The Evolution of Grandparents." *Scientific American* 305.2 (August 2011): 45–49.

Cassirer, Ernst. *The Philosophy of Symbolic Forms: Mythical Thought*. Translated by Ralph Manheim. New Haven, CT: Yale University Press, 1955.

Chang, Kwang-chih. "China on the Eve of the Historical Period." In *Cambridge History of Ancient China*, ed. Michael Loewe and Edward L. Shaughnessy. Cambridge: Cambridge University Press, 1999.

Chippendale, Christopher. *Stonehenge Complete*. Third ed. London: Thames & Hudson, 2004.

Christ, Carol P. "Mircea Eliade and The Feminist Paradigm Shift." *Journal of Feminist Studies in Religion* 7.2 (Fall, 1991): 75–94.

Clausen, Constance. "Creation by Sound / Creation by Light: A Sensory Analysis of Two South American Cosmologies." *The Varieties of Sensory Experience: A Sourcebook*, ed. David Howes. Toronto: University of Toronto Press, 1991.

Codrington, Robert Henry. "Religious Beliefs and Practices in Melanesia." *The Journal of the Anthropological Institute of Great Britain and Ireland* 10 (1881): 271–72.

Cooney, Gabriel. "Newgrange: A View from the Platform." *Antiquity* 80.309 (September 2006): 697–708.

Corless, Roger. "After Eliade, What?" Review of Mircea Eliade, *Essential Sacred Writings from Around the World*." *Religion* 23.4 (1993): 373–77.

Darvill, Timothy. "White on Blonde: Quartz Pebbles and the Use of Quartz at Neolithic Monuments in the Isle of Man and Beyond." In *Colouring the Past*, ed. Andrew Jones and Gavin MacGregor. Oxford: Berg Publishers, 2002.

Eliade, Mircea. *Cosmos and History*. New York: Harper & Row, 1959.

———. *Patterns in Comparative Religion*. Translated by Rosemary Sheed. New York: New American Library, 1958.

———. "Spirit, Light, and Seed," *History of Religions* 11.1 (August 1971): 1–30.

———. *The Two and the One*. Translated by J. M. Cohen. New York: Harper, 1965.

Ellwood, Robert E. *The Politics of Myth: A Study of C.G. Jung, Mircea Eliade, and Joseph Campbell*. Albany, NY: State University of New York Press, 1999.

Eriksen, Palle. "The Great Mound of Newgrange: An Irish Multi–Period Mound Spanning from the Megalithic Tomb Period to the Early Bronze Age." *Acta Archaeologica* 79 (2008): 250–73.

Fagan, Brian. *Cro-Magnon: How the Ice Age Gave Birth to the First Modern Humans*. New York: Bloomsbury Press, 2010.

Giedion, S. *The Eternal Present: The Beginnings of Architecture*. Mellon Lectures XXXV, Part 2. New York: Pantheon. 1964.

Haberland, Wolfgang. *The Art of North America*. New York: Crown, 1964.

Hadingham, Evan. *Early Man and the Cosmos*. Norman, OK: University of Oklahoma Press, 1984.

Herity, Michael. *Irish Passage Graves: Neolithic Tomb–Builders in Ireland and Britain 2500 B.C.* New York: Harper & Row, 1975.

Johnston, Susan A. Review, *Inside the Neolithic Mind: Consciousness, Cosmos, and the Realm of the Gods* by David Lewis-Williams and David Pearce. *American Anthropologist* 108.3 (September 2006): 600–01.

Jolly, Pieter. "On the Definition of Shamanism." *Current Anthropology* 46.1 (February 2005): 127–28.

Köhler, Walter. *Lighting in Architecture: Light and Color as Stereoplastic Elements*. New York: Reinhold, 1957.

Kramer, S. N., ed. *Mythologies of the Ancient World*. Garden City, NY: Doubleday, 1961.

Leach, Maria. *The Beginning*. New York: Funk and Wagnalls, 1956.

Lewis-Williams, David. *Inside the Neolithic Mind: Consciousness, Cosmos and the Realm of the Gods*. London: Thames & Hudson, 2005.

Lommel, Andreas. *Prehistoric and Primitive Man*. New York: McGraw–Hill, 1966.

Mackie, Euan W. "'Stonehenge: Neolithic Man and the Cosmos.' Review." *Antiquity* 71.273 (September 1997): 769–70.

McNeil, W. H. *The Rise of the West*. New York: NAL, 1965.

Michael, Henry N. *The Neolithic Age in Eastern Siberia*. Transactions of the American Philosophical Society N.S., v. 48, pt. 2. Philadelphia: The American Philosophical Society, 1958.

Mithen, Steven J. *After the Ice: A Global Human History, 20,000–5000 B.C.* Cambridge: Harvard University Press, 2004.

———, ed. *Creativity in Human Evolution and Prehistory*. London: Routledge, 1998.

Mohen, Jean-Pierre. *The World of Megaliths*. New York: Facts on File, 1990.

North, John. *Stonehenge: A New Interpretation of Prehistoric Man and the Cosmos*. New York: The Free Press, 1996.

O'Kelly, Michael J. *Early Ireland: An Introduction to Irish Prehistory*. Cambridge: Cambridge University Press, 1989.

Okladnikov, Alexei. *Art of The Amur: Ancient Art of the Russian Far East*. New York: Abrams, 1981.

Pearson, Mike Parker, et al. "The Age of Stonehenge." *Antiquity* 81.313 (September 2007): 617–39.

Pfaff, William. "Review of Francis Fukuyama, *The Origins of Political Order: From Prehuman Times to the French Revolution*." *The New York Review of Books* 58.18 (November 24, 2011): 71.

Ruggles, Clive. *Astronomy in Prehistoric Britain and Ireland*. New Haven, CT: Yale University Press, 1999.

———. "Interpreting Solstitial Alignments in Late Neolithic Wessex." *Archaeoastronomy* 20 (2006): 1–27.

Sandars, N. K. *Prehistoric Art in Europe*. Baltimore, MD: Penguin, 1968.

Sofaer, Anna, Volker Zinser, and Rolf M. Sinclair. "A Unique Solar Marking Construct." *Science* 206 (1979): 283–85.

Sproul, Barbara C. *Primal Myths; Creating the World*. New York: Harper & Row, 1979.

Stout, Geraldine. *Newgrange and the Bend of the Boyne*. Irish Rural Landscapes, vol. 1. Cork, Ireland: Cork University Press, 2002.

Tait, Charles. "About Maeshowe." Accessed August 9, 2011. http://www.maeshowe.co.uk/maeshoweabout.html.

Thompson, Stith. *Tales of the North American Indians*. New York: Midland, 1957 [1929].

———, ed. *Motif-Index of Folk-Literature*. Rev. and enl. Bloomington: Indiana University Press, 1966.

Twohig, Elizabeth Shee. *The Megalithic Art of Western Europe*. Oxford: Clarendon Press, 1981.

Vasseleu, Cathryn. *Textures of Light: Vision and Touch in Irigaray, Levinas and Merleau–Ponty*. London: Routledge, 1998.

Wald, Robert M. *Space, Time, and Gravity*. Chicago: University of Chicago Press, 1977.

Yourgrau, Wolfgang, and Allen D. Breck, eds. *Cosmology, History, and Theology*. New York: Plenum Press, 1977.

CHAPTER 3: EGYPTIAN CLASSICAL AGE

Aldred, Cyril. *The Development of Ancient Egyptian Art from 3200 to 1315 B. C.* London, A. Tiranti, 1962.

Allen, James P. "The Cosmology of the Pyramid Texts." *Religion and Philosophy in Ancient Egypt*, ed., William Kelly Simpson. New Haven, CT: Yale University Press, 1989.

———. "The Natural Philosophy of Akhenaten." *Religion and Philosophy in Ancient Egypt*, ed., William Kelly Simpson. New Haven, CT: Yale University Press, 1989.

———. "Shadow" *The Oxford Encyclopedia of Ancient Egypt*. Edited by Donald B. Redford. Accessed April 25, 2012. http://www.oxford–ancientegypt.com/entry?entry=t176.e0663.

Anthes, Rudolf. "Mythology in Ancient Egypt." *Mythologies of the Ancient World*, ed., Samuel Noah Kramer. New York: Doubleday, 1961.

Arts and Humanities Through the Ages, Vol. 1: *Ancient Egypt 2675-3323 B.C.E.*, ed. Edward Bleibert. New York: Gale, 2005.

Assmann, Jan. "Akhanyati's Theology of Light and Time." *Proceedings of the Israel Academy of Sciences and Humanities* 7.4 (1992): 143–76.

———. *Egyptian Solar Religion in the New Kingdom: Re, Amun and the Crisis of Polytheism*. Translated by Anthony Alcock. London: Kegan Paul International, 1995.

Badawy, Alexander. *A History of Ancient Architecture*. Berkeley: University of California Press, 1966.
Bell, Barbara. "Climate and the History of Egypt: The Middle Kingdom." *American Journal of Archaeology* 79.3 (1975): 223–69.
Bose, Abinash Chandra. *Hymns from the Vedas*. Bombay: Asia Publishing House, 1966.
Breasted, James Henry, ed. and trans. *Ancient Records of Egypt*. New York: Russell & Russell, 1962.
Campbell, Joseph. *The Masks of God: Oriental Mythology*. New York: Viking, 1962.
Cruikshank, William. "Light and Darkness." In *Hastings Encyclopedia of Religion and Ethics*, ed. James Hastings and J. A. Selbie. 13 vols. New York: Scribner's, 1909–1926.
David, Rosalie. *The Experience of Ancient Egypt*. New York: Routledge, 2000.
De Cenival, Jean-Louis. *Living Architecture: Egyptian*. New York: Grosset & Dunlap, 1964.
Doxey, Denise M. "Priesthood." *The Oxford Encyclopedia of Ancient Egypt*. Edited by Donald B. Redford. Accessed May 23, 2012. http://www.oxford–ancientegypt.com/entry?entry=t176.e0581.
Dresden, M. J. "Mythology of Ancient Iran." *Mythologies of the Ancient World*, ed., Samuel Noah Kramer. New York: Doubleday, 1961.
Edwards, I. E. S. *The Pyramids of Egypt*. New York: Viking, 1985.
The Egyptian Book of the Dead. Edited by Carol A. R. Andrews. Translated by Raymond O. Faulkner and Ogden Goelet, Jr. Rev. ed. San Francisco: Chronicle Books, 2008.
The Electronic Corpus of Sumerian Literature, Copyright 2003, 2004, 2005, 2006. The ETCSL project, Faculty of Oriental Studies, University of Oxford, http://etcsl.orinst.ox.ac.uk/.
Eliade, Mircea. *Patterns in Comparative Religion*. Translated by Rosemary Sheed. New York: New American Library, 1958.
Encyclopedia of Religion. 15 vols. Second ed. Edited by Lindsay Jones. New York: Macmillan Reference, 2005.
Erman, Adolf. *The Literature of the Ancient Egyptians*. Translated by Aylward M. Blackman. New York: Benjamin Blom, 1971 [1927].
Fakhry, Ahmed. *The Pyramids*. Chicago: University of Chicago Press, 1961.
Faulkner, R. O., trans. *The Ancient Egyptian Coffin Texts*. 3 vols. London: Warminster, 1973–1978.
———, trans. *The Egyptian Book of the Dead: The book of Going Forth by Day*. San Francisco: Chronicle Books, 1995.
Foster, John L. *Hymns, Prayers, and Songs: An Anthology of Ancient Egyptian Lyric Verse*. Atlanta: Society of Biblical Literature, 1995.
Freed, R. Y. Markowitz, and S. D'Auria, eds. *Pharaohs of the Sun: Akhenaten, Nefertiti, Tutankhamen*. Boston, 1999.
Giedion, S. *The Eternal Present: The Beginnings of Architecture*. Mellon Lectures XXXV, Part 2. New York: Pantheon. 1964.
Goelet, Ogden. Introduction to *The Egyptian Book of the Dead: The Book of Going Forth by Day*. Translated by Raymond Faulkner. San Francisco: Chronicle Books, 1995.
Gray, Louis Herbert, ed. *The Mythology of All Races*. 13 vols. New York: Cooper Square, 1964.
Gundlach, Rolf, and Robert E. Shillenn. "Temples." *The Oxford Encyclopedia of Ancient Egypt*. Edited by Donald B. Redford. Accessed April 25, 2012. http://www.oxford–ancientegypt.com/entry?entry=t176.e0711.
Hari, Robert. *New Kingdom Amarna Period*. Institute of Religious Iconography State University of Groningen. Leiden: E. J. Brill, 1985.
Hastings Encyclopedia of Religion and Ethics. Eds., James Hastings and J. A. Selbie. 13 vols. New York: Scribner's, 1909–1926.
Hollis, Susan. Review of *Akhenaten and the Religion of Light* by Erik Hornung. *Journal of the American Oriental Society* 121.3 (July 2001): 500.
Hornung, Erik. *Akhenaten and the Religion of Light*. Translated by David Lorton. Ithaca, NY: Cornell University Press, 1999.
———. *Idea into Image: Essays on Ancient Egyptian Thought*. Translated by Elizabeth Bredeck. New York, 1992.

Kemp, Barry. *Ancient Egypt: Anatomy of a Civilization*. Second ed. New York: Routledge, 2006 [1989].

Kostof, Spiro. *A History of Architecture: Settings and Rituals*. New York: Oxford University Press, 1985.

Kovacs, Maureen Gallery, trans. and ed. *The Epic of Gilgamesh*. Sanford, CA: Stanford University Press, 1989.

Lacovara, Peter. "The City of Amarna." In *Pharaohs of the Sun: Akhenaten, Nefertiti, Tutankhamen*, ed. Rita Freed, et al. Boston: Thames & Hudson, 1999.

Lehner, Mark. *The Complete Pyramids*. London: Thames and Hudson, 1997.

Lesko, Leonard H. "Book of Going Forth By Day." *The Oxford Encyclopedia of Ancient Egypt*. Edited by Donald B. Redford. Accessed June 8, 2012. http://www.oxford-ancientegypt.com/entry?entry=t176.e0099.

Levin, Kim. "The Eye of Ra." *Light: From Aten to Laser*, eds. Thomas B. Hess and John Ashbery. Art News Annual XXXV. New York: Macmillan, 1969.

Lichtheim, Miriam. *Ancient Egyptian Literature*. 3 vols. Berkeley, CA: University of California Press, 1973–1980.

Lloyd, Alan B., ed. *A Companion to Ancient Egypt*. 2 vols. Malden, MA: Wiley–Blackwell, 2010.

Loprieno, Antonio. *Ancient Egyptian Literature*. Leiden, 1996.

Manuelian, Peter Der, ed. *The Ancient Egyptian Pyramid Texts*. Translated by James P. Allen. Atlanta: Society of Biblical Literature, 2005.

Mason, Peter. *Light Fantastic*. London: Penguin, 1981.

Michalowski, Kazimierz. *Art of Ancient Egypt*. Translated by Norbert Guterman. New York: Harry N. Abrams, n.d.

Morenz, Siegfried. *Egyptian Religion*. Translated by Ann E. Keep. Ithaca: Cornell University Press, 1992.

Nissen, Hans Jörg. *The Early History of the Ancient Near East, 9000-2000 B. C.* Translated by Elizabeth Lutzeier. Chicago: University of Chicago Press, 1988.

Parcak, Sarah. "The Physical Context of Ancient Egypt." In *A Companion to Ancient Egypt*, ed. by Alan B. Lloyd. 2 vols. Malden, MA: Wiley–Blackwell, 2010.

Plas, Dirk Van Der, John L. Foster, and James P. Allen. "Hymns." *The Oxford Encyclopedia of Ancient Egypt*. Edited by Donald B. Redford. Accessed April 25, 2012. http://www.oxford ancientegypt.com/entry?entry=t176.e0329.s0003.

Poo, Mu-chou. Review of *Egyptian Solar Religion in the New Kingdom* by Jan Assmann. *Journal of the Economic and Social History of the Orient* 40.3 (1997): 299–300.

Quirke, Stephen. *The Cult of Ra: Sun–Worship in Ancient Egypt*. New York: Thames & Hudson, 2001.

Rad, Gerhard von. *Genesis*. Philadelphia: The Westminster Press, 1961.

Redford, Donald B. *Akhenaten: The Heretic King*. Princeton, NJ: Princeton University Press, 1984.

———. *Egypt, Canaan, and Israel in Ancient Times*. Princeton, NJ: Princeton University Press, 1992.

———, ed. *The Oxford Encyclopedia of Ancient Egypt*. 3 vols. E-reference edition. New York: Oxford University Press, 2001, 2005.

———. "The Sun-disc in Akhenaten's Program: Its Worship and Its Antecedents." *Journal of the American Research Center in Egypt* 13 (1976): 47–61; 17 (1982): 21–38.

Robins, Gay. *The Art of Ancient Egypt*. Cambridge, MA: Harvard University Press, 1997.

———. "Color Symbolism." *The Oxford Encyclopedia of Ancient Egypt*. Edited by Donald B. Redford. Accessed April 25, 2012. http://www.oxford–ancientegypt.com/entry?entry=t176.e0147.

Romano, James F. "Solar Culture in Ancient Egypt." In *The Sun: Symbol of Power and Life*, ed. Madanjeet Singh. New York: H.N. Abrams: Unesco, 1993.

Russman, Edna. *Egyptian Sculpture. Cairo and Luxor*. Austin: University of Texas Press, 1989.

Schäfer, Heinrich. *Principles of Egyptian Art*. Oxford: Clarendon Press, 1974.

Schlögl, Hermann A. "Aten." *The Ancient Gods Speak: A Guide to Egyptian Religion*. Ed. Donal B. Redford. New York: Oxford University Press, 2002.
Shafer, Byron E., ed. *Religion in Ancient Egypt: Gods, Myths, and Personal Practice*. Ithaca, NY: Cornell University Press, 1991.
Shepperson, Mary. "Planning for the Sun: Urban Forms as a Mesopotamian Response to the Sun." *World Archaeology* 41.3 (2009): 363–78.
Sibum, L. *The Bible on Light*. Translated by F. Vander Heijden. De Pere, WS: St. Norbert Abbey Press, 1966.
Simpson, William Kelly, ed. *The Literature of Ancient Egypt*. Translated by R. O. Faulkner, et al. New Haven: Yale University Press, 1972.
Singh, Madanjeet. *The Sun: Symbol of Power and Life*. New York: H.N. Abrams: Unesco, 1993.
Smith, William Stevenson. *The Art and Architecture of Ancient Egypt*. Baltimore, MD: Penguin, 1965.
———. "Egyptian Art." In *Readings in Art History*, ed. Harold Spencer. Vol 1. Third ed. New York: Charles Scribner's Sons, 1969.
Spencer, Neal. "Priests and Temples: Pharaonic." *A Companion to Ancient Egypt*, ed., Alan B. Lloyd. Vol. 2. Malden, MA: Wiley-Blackwell, 2010.
Stewart, H. M. "Traditional Egyptian Sun Hymns of the New Kingdom." *Bulletin of the Institute of Archaeology, University of London* 6 (1967): 29–74.
Szpakowska, Kasia. "Religion in Society: Pharaonic." *A Companion to Ancient Egypt*, ed., Alan B Lloyd. 2 vols. Malden, MA: Wiley–Blackwell, 2010.
Thomas, Nancy. "Petamenophis." *The Oxford Encyclopedia of Ancient Egypt*. Edited by Donald B. Redford. Accessed May 23, 2012. http://www.oxford–ancientegypt.com/entry?entry=t176.e0581.
Trigger, Bruce G. "Review of *Egyptian Solar Religion in the New Kingdom* by Jan Assmann." *The International Journal of African Historical Studies* 29.2 (1996): 424–26.
Troy, Lana. "Review of *Egyptian Solar Religion in the New Kingdom* by Jan Assmann." *The African Archaeological Review* 13.2 (June 1996): 153–163.
Wilkinson, Toby A. H. *The Rise and Fall of Ancient Egypt*. New York: Random House, 2010.
Wilson, John A. *The Culture of Ancient Egypt*. Chicago: University of Chicago Press, 1951.
Wilson, Penelope. "Temple Architecture and Decorative Systems." *A Companion to Ancient Egypt*, ed. Alan B. Lloyd. Vol 2. Malden, MA: Wiley–Blackwell, 2010.
Winter, I. "Radiance as an Aesthetic Value in the Art of Mesopotamia." In *Art: The Integral Vision*, ed. B. N. Saraswati, S. C. Malik, and M. Khanna. New Delhi: D. K. Printworld, 1994.
Wolde, Ellen van. " In Words and Pictures: The Sun in 2 Samuel 12:7-12." *Biblical Interpretation: A Journal of Contemporary Approaches* 11.3-4 (2003): 259–78.
Wolf, W. *The Origins of Western Art: Egypt, Mesopotamia, the Aegean*. New York: Universe Books, 1989.
Zandee, J. "Prayers to the Sun–God from Theban Tombs." *Jaarbericht van het Vooraziatisch–Egyptische Genootschap "Ex Oriente Lux"* 16 (1959–1962): 48–71.
———. "Hymnical Sayings Addressed to the Sun–God by the High–Priest Nebwenenef, from his Tomb in Thebes." *Jaarbericht van het Vooraziatisch–Egyptische Genootschap "Ex Oriente Lux"* 18 (1964): 253–65.

CHAPTER 4: GREEK CLASSICAL AGE

Aeschylus. *The Oresteian Trilogy*. Translated by Philip Vellacott. Baltimore: Penguin Books, 1969.
Agrelo, Santiago. "El Tema Biblíco de la Luz." *Antonianum* 50 (1975): 353–417.
Andrewes, Antony. *The Greeks*. New York: Knopf, 1967.
Armstrong, Karen. *The Great Transformation*. New York: Anchor, 2006.
Ascherl, Jolana. *Das Licht in der Pompejanischen Wandermalerei*. 2 vols. Regensbury: S. Roderer, 2002.

Authier, Michel. "Refraction and Cartesian Forgetfulness." *A History of Scientific Thought: Elements of a History of Science*, ed. Michel Serres. Cambridge, MA: Blackwell, 1995.

Barlow, Shirley A. *The Imagery of Euripides*. London: Methuen, 1971.

Beard, Mary. *The Parthenon*. Cambridge, MA: Harvard University Press, 2010.

———. "Adopting an Approach." *Looking at Greek Vases*, ed. Tom Rasmussen and Nigel Spivey. New York: Cambridge University Press, 1991, 12–35.

———. "'The Latest Scheme for the Parthenon,' review of *The Parthenon Enigma* by Joan Breton Connelly (New York: Knopf, 2014)." *The New York Review of Books* LXI.4 (March 6, 2014): 14–16.

Bellah, Robert N. *Religion in Human Evolution: From the Paleolithic to the Axial Age*. Cambridge: Harvard University Press, 2011.

———, and Hans Joas, eds. *The Axial Age and Its Consequences*. Cambridge, MA: Harvard University Press, 2012.

Bernidaki-Aldous, Eleftheria A. *Blindness in a Culture of Light*. New York: Peter Lang. 1990.

Berryman, Sylvia. "Euclid and the Skeptic: A Paper on Vision, Doubt, Geometry, Light and Drunkenness." *Phronesis* 43.2 (May 1998): 176–96.

Blumenberg, Hans. "Light as a Metaphor for Truth." *Modernity and the Hegemony of Vision*, ed. David Michael Levin. Berkeley, CA: University of California Press, 1993.

Boardman, John. *The Parthenon and its Sculptures*. Austin: University of Texas, 1985.

———. *Athenian Black Figure Vases: A Handbook*. New York: Thames & Hudson, 1985.

———. *Athenian Red Figure Vases: The Classical Period: A Handbook*. New York: Thames & Hudson, 1989.

Borzacchini, L. "Light as a Metaphor for Science." *Semiotica* 136.1/4 (2001): 151–71.

Bose, Abinash Chandra, ed. *Hymns from the Vedas*. Translated by A. C. Bose. New York: Asia Publishing House, 1966.

Boyce, P. R. "The Luminous Environment." *Environmental Interaction: Psychological Approaches to Our Physical Surroundings*, ed. D. Cantor and P. Stringer. New York: International Universities Press, 1975, 81–124.

Brilliant, Richard. *Arts of the Ancient Greeks*. New York: McGraw-Hill, 1973.

Bruno, Vincent J. *Form and Color in Greek Painting*. New York: Norton, 1977.

Burnet, John. *Early Greek Philosophy*. New York: Meridian, 1965.

Calasso, Roberto. *The Marriage of Cadmus and Harmony*. New York: Knopf, 1993.

Cantor, G. N. "Physical Optics." *Companion to the History of Modern Science*, ed. R. C. Olby, et al. London: Routledge, 1990.

Carpenter, Rhys. "The Aesthetics of Greek Architecture." *Readings in Art History*, ed. Harold Spencer. New York: Charles Scribner's Sons, 1982. I: 111–27.

———. "Rev. of *Art and Experience in Classical Greece*, by J.J. Pollitt." *American Journal of Archeology* 77 (1972): 349.

Charbonneaux, Jean, Roland Martin, and François Villard, eds. *Classical Greek Art (480-330 B.C.)*. Translated by James Emmons. New York: George Braziller, 1972.

———. *Hellenistic Art (330-50 B.C.)*. Translated by Peter Green. New York: George Braziller, 1973.

Chidester, David. *Word and Light: Seeing, Hearing and Religious Discourse*. Chicago: University of Illinois Press, 1992.

Cohen, Morris R., and I. E. Drabkin, eds. *A Source Book in Greek Science*. New York: McGraw–Hill, 1948.

Collis, John Stewart. *The World of Light*. New York: Horizon Press, 1960.

A Companion to the Hellenistic World. Edited by Andrew Erskine. Malden, MA: Blackwell, 2003.

Constantinidou, Soteroula. "The Light Imagery of Divine Manifestation in Homer." In *Light and Darkness in Ancient Greek Myth and Religion*, ed. Christopoulos, Menelaos, et al. New York: Rowman & Littlefield, Lexington, 2010, 91–109.

Cook, Robert Manuel. *Greek Art: Its Development, Character and Influence*. New York: Farrar, 1973.

Copleston, Frederick. *A History of Philosophy: Greece and Rome*. New revised ed. New York: Image Books, 1962.

Cuttle, Christopher. "The Sharpness and the Flow of Light." In *Architectural Psychology: Proceedings of the Lund Conference*, ed. R. Küller. Stroudsberg, Germany: Dowden, Huthinson & Ross, 1973, 12–22.

Davies, Malcolm. "'Light and Darkness.' Rev. of *Licht und Dunkel in der frühgriechischen Dichtung*, by Dieter Bremer." *The Classical Review* (New Series) 29.1 (1979): 1–3.

Derrida, Jacques. "White Mythology." *New Literary History* 6.1 (1974): 5–74.

A Dictionary of Greek and Roman Geography. 2 vols. Edited by William Smith. New York: AMS Press, 1966.

Dictionary of Untranslatables: A Philosophical Lexicon. Edited by Barbara Cassin. Translated by Steven Rendall, et al. Princeton, NJ: Princeton University Press, 2014.

Dionysius the Areopagite. *On the Divine Names, and, Mystical Theology*. Translated by C. E. Rolt. New York: Macmillan, 1951.

Donald, Merlin. "Art and Cognitive Evolution." In *The Artful Mind: Cognitive Science and the Riddle of Human Creativity*, ed. Mark Turner. New York: Oxford University Press, 2006, 3–20.

Doniger, Wendy. *The Rig Veda: An Anthology*. New York: Penguin, 1981.

Eliade, Mircea. "Spirit, Light, and Seed." *History of Religions* 11.1 (1971): 1–30.

———. *The Two and the One*. Translated by J. M. Cohen. Chicago: University of Chicago Press, 1962.

Euripides. *The Bacchae*. Translated by Geoffrey S. Kirk. Englewood Cliffs, NJ: Prentice-Hall, 1970.

Everson, Stephen. *Aristotle on Perception*. New York: Oxford University Press, 1997.

Fontaine, Petrus F. M. *The Light and the Dark: A Cultural History of Dualism*. 17 vols. Amsterdam, Netherlands: J. C. Gieben, 1988.

Fox, Robin Lane. *The Classical World*. New York: Basic Books, 2006.

Fränkel, Hermann. *Early Greek Poetry and Philosophy*. Translated by Moses Hadas and James Willis. New York: Harcourt Brace Jovanovich, 1973.

Frankfurt, Henri. "Conclusion: The Emancipation of Thought from Myth." In *The Intellectual Adventure of Ancient Man: An Essay of Speculative Thought in the Ancient Near East*, ed. Henri Frankfort, et al. Chicago: University of Chicago Press, 1977.

Gombrich, E. H. *Art and Illusion*. Second ed. New York: Pantheon, 1961.

———. "The Heritage of Apelles." In *The Heritage of Apelles: Studies in the Art of the Renaissance*. Ithaca, NY: Cornell University Press, 1976.

The Grove Encyclopedia of Classical Art and Architecture. Edited by Gordon Campbell. E-reference, Monmouth University, 2007.

Guthrie, W. K. C. *A History of Greek Philosophy*. 3 vols. Cambridge: Cambridge University Press, 1965.

Hall, Edith. *The Theatrical Cast of Athens*. Oxford: Oxford University Press, 2006.

Haney, William S. "Metaphor and the Experience of Light." *Analecta Husserliana* 38 (1992): 237–38.

Hanfmann, George M. A. "Hellenistic Art." In *Readings in Art History*, ed. Harold Spencer. New York: Charles Scribner's Sons, 1982, I: 129–47.

Herman, Arthur. *The Cave and the Light: Plato Versus Aristotle, and the Struggle for the Soul of Western Civilization*. New York: Random House, 2013.

Hermann, Arnold. *To Think Like God: Pythagoras and Parmenides, the Origins of Philosophy*. Las Vegas, NV: Parmenides Publishing, 2004.

Holt, Philip. "Light in Sophokles' *Trachiniai*." *Classical Antiquity* 6.2 (October 1987): 205–17.

Hornblower, Simon. *The Greek World: 479–323 BC*. Fourth ed. New York: Routledge, 2011.

Hughes, J. Donald. "The Natural Environment." In *A Companion to the Classical Greek World*, ed. Konrad Kinzl. Malden, MA: Blackwell, 2006.

Johnston, Sarah Iles. "*Fiat Lux, Fiat Ritus*: Divine Light and the Late Antique Defense of Ritual." In *The Presence of Light: Divine Radiance and Religious Experience*, ed. Matthew T. Kapstein. Chicago: University of Chicago Press, 2004, 5–24.

Kahler, Erich. "The Nature of the Symbol." In *Symbolism in Religion and Literature*, ed. Rollo May. New York: George Braziller, 1960.

Kearney, Richard. *Poetics of Imagining: From Husserl to Lyotard*. New York: HarperCollins, 1991.

Kearns, Emily. "Religious Practice and Belief." In *A Companion to the Classical Greek World*, ed. Konrad Kinzl. Malden, MA: Blackwell, 2006.

Kerényi, Carl. *Dionysos: Archetypal Image of Indestructible Life*. Translated by Ralph Manheim. Princeton: Princeton University Press, 1976.

Kinzl, Konrad, ed. *A Companion to the Classical Greek World*. Malden, MA: Blackwell, 2006.

Knox, Bernard, ed. *The Norton Book of Classical Literature*. New York: W.W. Norton, 1993.

Lattimore, Steven. "From Classical to Hellenistic Art." In *A Companion to the Classical Greek World*, ed. Konrad Kinzl. Malden, MA: Blackwell, 2006.

Lawrence, Arnold W. *Greek Architecture*. Fourth ed. Revised with additions by R. A. Tomlinson. New York: Penguin, 1983.

Levi, Peter. "Greek Drama." In *The Oxford History of the Classical World*, ed. John Boardman, et al. Oxford: Oxford University Press, 1986, 156–85.

Lindberg, *Theories of Vision from Al–Kindi to Kepler*. Chicago: University of Chicago Press, 1976.

Marchetti, Laura. "Light as an Original Metaphor." *Semiotica* 136.1/4 (2001): 245–68.

Marinatos, Nanno. "Light and Darkness and Archaic Greek Cosmography." In *Light and Darkness in Ancient Greek Myth and Religion*, ed. Christopoulos, Menelaos, et al. New York: Rowman & Littlefield, Lexington, 2010, 91–109, 193–99.

Mark, Robert. *Light, Wind, and Structure*. Cambridge, MA: MIT Press, 1990.

McClain, Ernest G. *The Myth of Invariance: The Origins of the Gods, Mathematics and Music from the Ṛg Veda to Plato*. New York: Nicolas Hays, 1976.

McEvilley, Thomas. *The Shape of Ancient Thought: Comparative Studies in Greek and Indian Philosophies*. New York: Allworth Press, 2002.

Mehta, Rohit. *The Call of the Upanishads*. Delhi: Motilal Barnarsidass, 1970.

Menelaos, Christopoulos, et al, eds. *Light and Darkness in Ancient Greek Myth and Religion*. New York: Rowman & Littlefield, Lexington, 2010.

Mertens, Joan R. *How to Read Greek Vases*. New York: Metropolitan Museum of Art, 2010.

Milleker, Elizabeth J., and Joseph Coscia, Jr. *Light on Stone, Greek and Roman Sculpture in the Metropolitan Museum of Art: A Photographic Essay*. New Haven, CT: Yale University Press, 2003.

Murphy, N. R. "The 'Simile of Light' in Plato's *Republic*." *The Classical Quarterly* 26.2 (April 1932: 93–102.

Needham, Joseph. *Science and Civilization in China*. 5 vols. Cambridge: Cambridge University Press, 1988.

Nikhilananda, Swami, ed. *The Upanishads*. New York: Harper, 1964.

Notopoulos, J. A. "The Symbolism of the Sun and Light in the *Republic* of Plato." *Classical Philology* 39.3 (July 1944): 163–72; 39.4 (October 1944): 223–40.

O'Grady, Patricia F. *Thales of Miletus: The Beginnings of Western Science and Philosophy*. Aldershot, UK: Ashgate 2002.

The Oxford Classical Dictionary. Third ed. Edited by Simon Hornblower and Antony Spawforth. New York: Oxford University Press, 1996.

The Oxford Dictionary of Classical Myth and Religion. Edited by Simon Price and Emily Kearns. Oxford: Oxford University Press, 2003.

Oxford Encyclopedia of Ancient Greece and Rome. 7 vols. Edited by Michael Gagarin. New York: Oxford University Press, 2010.

The Oxford History of Classical Art. Edited by John Boardman. New York: Oxford University Press, 1993.

Paleothodoros, Dimitris. "Light and Darkness in Dionysiac Rituals, as Illustrated on Attic Vase Paintings of the 5 th Century BCE." In *Light and Darkness in Ancient Greek Myth and Religion*, ed. Christopoulos, Menelaos, et al. New York: Rowman & Littlefield, Lexington, 2010, 237–60.

Palmer, John. "Parmenides." *Stanford Encyclopedia of Philosophy*. Summer 2012 edition. Edited by Edward N. Zalta. http://plato.stanford.edu/archives/sum2012/entries/parmenides/.

Pappas, Nickolas. *Routledge Philosophy Guidebook to Plato and the Republic*. Second ed. New York: Routledge, 1995.

Parisinou, Eva. *The Light of the Gods: The Role of Light in Archaic and Classical Greek Cult*. London: Duckworth, 2000.

Park, David. *The Fire within the Eye*. Princeton, NJ: Princeton University Press, 1997.

Paul, Judith Thomson. "The Propitious and the Perverse: Light and Dark Imagery in the Oresteia." Diss., Cornell University, 1981.

Penglase, Charles. *Greek Myths and Mesopotamia: Parallels and Influence in the Homeric Hymns and Hesiod*. New York: Routledge, 1994.

Peradotto, John J. "Some Patterns of Nature Imagery in the *Oresteia*." *The American Journal of Philology* 85.4 (October 1964): 378–93.

Plato. *Plato Complete Works*. Edited by John M. Cooper. New York: Hackett, 1997.

———. *Plato: Collected Dialogues*. Edited by Edith Hamilton and Huntington Cairns. Princeton, NJ: Princeton University Press, 1973.

Pollitt, J. J. *The Ancient View of Greek Art: Criticism, History, and Terminology*. New Haven, CT: Yale University Press, 1974.

———. *Art and Experience in Classical Greece*. Cambridge: Cambridge University Press, 1972.

———. *The Art of Greece 1400-31 B.C.: Sources and Documents*. Englewood Cliffs, NJ: Prentice-Hall, 1965.

Ponzio, J. "Light, Being, and Time." *Semiotica* 136.1-4 (2001): 55–67.

Popper, K. R. "How the Moon Might Shed some of Her Light Upon the Two Ways of Parmenides." *Classical Quarterly* 42 (1992): 12–19.

———. *The World of Parmenides: Essays on the Presocratic Enlightenment*. New York: Routledge, 1998.

Price, Simon. *Religions of the Ancient Greeks*. Cambridge: Cambridge University Press, 1999.

Rasmussen, Tom, and Nigel Spivey, eds. *Looking at Geek Vases*. New York: Cambridge University Press, 1991.

Richter, Gisela M. A. *Attic Red-figured Vases*. New Haven, CT: Yale University Press, 1958.

———. "The General Characteristics of Greek Sculpture." In *Readings in Art History*, ed. Harold Spencer. New York: Charles Scribner's Sons, 1969, I: 37–49.

Robertson, Martin. *A History of Greek Art*. 2 vols. Cambridge: Cambridge University Press 1975.

———. *Greek Painting*. New York: Rizzoli International Publications, 1979

———. "Adopting an Approach." In *Looking at Greek Vases*, ed. Tom Rasmussen and Nigel Spivey. New York: Cambridge University Press, 1991, 1–12.

Russo, Nicholas Mark. "The Imagery of Light and Darkness in the Oresteia." PhD. diss. Ohio State University, 1974.

Saunders, Jason L., ed. *Greek and Roman Philosophy After Aristotle*. New York: Free Press, 1966.

Schaps, David M. *Handbook for Classical Research*. New York: Routledge, 2011.

Sedley, David. "Parmenides." *Routledge Encyclopedia of Philosophy*. London: Routledge, 1998.

Serres, Michel. *The Birth of Physics*. Translated by J. Hawkes. Edited by D. Webb. Manchester: Clinamen, 2000.

———. "Gnomon: The Beginnings of Geometry in Greece." In *A History of Scientific Thought: Elements of a History of Science*, ed. Michel Serres. Cambridge, MA: Blackwell, 1995.

Silverman, A. "Colour and Colour–Perception in Aristotle's *de Anima*." *American Philosophical Quarterly* 9 (1989): 271–92.

Silvestri, Domenico. "From the Eloquence of Light to the Splendor of the Word." *Semiotica* 136.1/4 (2001): 117–32.

Snyder, Laura J. "'A Man Who Shaped Our World.' Rev. of *The King of Infinite Space*, by David Berlinski." *The Wall Street Journal* (February 23–24, 2013): C8.

Starkey, Jennifer S. "Play in the Sunshine." *Didaskalia* 8.22 (2011). Accessed March 5, 2013. www.didaskalia.net/issues/8/22.

Steele, Laura D. "Mesopotamian Elements in the Proem of Parmenides?" *The Classical Quarterly* (New Series) 52.02 (2002): 583–88.

Steinberger, Peter J. *The Problem With God*. New York: Columbia University Press, 2013.

Stewart, Andrew F. *Classical Greece and the Birth of Western Art*. Cambridge: Cambridge University Press, 2008.

———. *Greek Sculpture*. New Haven: Yale University Press, 1990.

Summers, David. *Vision, Reflection, and Desire in Western Painting*. Chapel Hill: University of North Carolina Press, 2007.

———. *Real Spaces*. New York: Phaidon, 2003.

Syropoulos, Spyros. "S-light Anomaly: Dark Brightness in Euripides' *Medea*." In *Light and Darkness in Ancient Greek Myth and Religion*, ed. Menelaos Christopoulos, et al. New York: Rowman & Littlefield, Lexington, 2010, 77–90.

Taplin, Oliver. *Greek Tragedy in Action*. Berkeley: University of California Press, 1978.

———. *Pots and Plays: Interactions Between Tragedy and Greek Vase-Painting of the Fourth Century BC*. Los Angeles: J. Paul Getty Museum, 2007.

Tarrant, Dorothy. "Greek Metaphors of Light." *Classical Quarterly* liv (1960): 181–87.

Tsitsibakou-Vasalos, Evanthia. "Brightness and Darkness in Pindar's *Pythian 3* Aigla–Koronis–Arsinoë and her Coming of Age." In *Light and Darkness in Ancient Greek Myth and Religion*, ed. Menelaos Christopoulos, et al. Plymouth, UK: Lexington Books, 2010.

Van Keuren, F. *Guide to Research in Classical Art and Mythology*. Chicago: American Library Association, 1991.

Vickers, Brian. *Towards Greek Tragedy: Drama, Myth, Society*. London: Longman, 1973.

Villard, François. "Landscape, Nature and Realism." In *Hellenistic Art (330-50 B.C.)*, ed. Jean Charbonneaux, et al. Translated by Peter Green. New York: George Braziller, 1973.

———. "Light and Colour (280–150)." In *Hellenistic Art (330–50 B.C.)*, ed. Jean Charbonneaux, et al. Translated by Peter Green. New York: George Braziller, 1973.

Viollet-Le-Duc, Eugène Emmanuel. *Lectures on Architecture*. 2 vols. Translated by Benjamin Bucknall. Boston: James R. Osgood, 1881.

Vlastos, Gregory. *Studies in Greek Philosophy I: The Presocractics*. Princeton: Princeton University Press, 1993.

Werblowsky, R. J. Zwi, and Julia Iwersen. "Light and Darkness." In *Encyclopedia of Religion*, second ed., ed. Lindsay Jones. New York: Macmillan, 2005.

Williams, Dyfri. "The Invention of the Red-figure Technique and the Race Between Vase-painting and Free Painting." In *Looking at Greek Vases*, ed. Tom Rasmussen and Nigel Spivey. New York: Cambridge University Press, 1991, 103–18.

Witzel, Michael. "Vedas and Upaniṣads." In *The Blackwell Companion to Hinduism*, ed. Gavin Flood. Malden, MA: Blackwell, 2003, 68–100.

Zajonc, Arthur. *Catching the Light: The Entwined History of Light and Mind*. New York: Bantam, 1993.

CHAPTER 5: ROMAN CLASSICAL AGE

Achtemeier, Elizabeth R. "Jesus Christ, the Light of the World: The Biblical Understanding of Light and Darkness." *Interpretation; a Journal of Bible and Theology* 17 (1963): 439–49.

Agrelo, Santiago. "El Tema Biblíco de la Luz." *Antonianum* 50 (1975): 353–417.

The Anchor Bible Dictionary. Edited by David Noel Freedman. 6 vols. New York: Doubleday, 1992.

Armstrong, A. H. "'Emanation' in Plotinus." *Mind* (New Series) 46.181 (January 1937): 61–66.

———. "Plotinus." *The Cambridge History of Later Greek and Early Medieval Philosophy*. 1967.

Ascherl, Jolana. *Das Licht in der Pompejanischen Wandermalerei*. 2 vols. Regensbury: S. Roderer, 2002.

Athanassiadi, P. "A Contribution to Mithraic Theology: The Emperor Julian's Hymn to King Helios." *The Journal of Theological Studies* (New Series) XXVIII (1977): 360–71.
Attridge, H. W. "Philo and John: Two Riffs on One Logos." *The Studia Philonica Annual* 17 (2005): 103–17.
Augustine of Hippo. *Confessions*. Translated by Henry Chadwick. New York: Oxford University Press, 1991.
Baker-Brian, Nicholas J. *Manichaeism: An Ancient Faith Rediscovered*. London: T&T Clark, 2011.
Barnstone, Willis, ed. *The Other Bible*. New York: Harper, 1984.
Blumenberg, Hans. "Light as a Metaphor for Truth." In *Modernity and the Hegemony of Vision*, ed. David Michael Levin. Berkeley, CA: University of California Press, 1993.
Borgen, Peder. *Logos was the True Light, and Other Essays on the Gospel of John*. Trondheim, Norway: Tapir, 1983.
Bowersock, G. W. *Julian the Apostate*. Cambridge: Harvard University Press, 1978.
Bréhier, Emile. "Philo Judaeus." *The Catholic Encyclopedia*. 1911. Web.
———. *The Philosophy of Plotinus*. Translated by J. Thomas. Chicago: University of Chicago Press, 1958.
Brown, Peter. *Through the Eye of the Needle: Wealth, the Fall of Rome, and the Making of Christianity in the West, 350-550 AD*. Princeton: Princeton University Press, 2012.
Bryson, Norman. *Looking at the Overlooked: Four Essays on Still Life Painting*. Cambridge, MA: Harvard University Press, 1990.
Carmichael, C. M. *The Story of Creation: Its Origins and Interpretation in Philo and the Fourth Gospel*. Ithaca, NY: Cornell University Press, 1996.
Chadwick, Henry. *Early Christian Thought and the Classical Tradition: Studies in Justin, Clement, and Origen*. Oxford: Oxford University Press, 1966.
———. "Philo." *The Cambridge History of Later Greek and Early Medieval Philosophy*. 1967.
Charbonneaux, Jean, Roland Martin, and François Villard, eds. *Hellenistic Art (330-50 B.C.)*. Translated by Peter Green. New York: George Braziller, 1973.
Chidester, David. *Word and Light: Seeing, Hearing and Religious Discourse*. Chicago: University of Illinois Press, 1992.
Copleston, Frederick. *A History of Philosophy: Greece and Rome*. New revised ed. New York: Image Books, 1962.
Corrigan, K. "Light and Metaphor in Plotinus and St. Thomas Aquinas." *The Thomist* 57.1 (1993): 187–99.
———. *Reading Plotinus: A Practical Introduction to Neoplatonism*. W. Lafayette, IN: Purdue University Press, 2005.
———. "El simbolismo natural de la luz en Plotino." *Revista de Filosofía* 25–26 (1985): 51–56.
Cumont, Franz Valery Marie. *The Mysteries of Mithra*. Translated by Thomas J. McCormack. New York: Dover, 1956.
Davies, Malcolm. "'Light and Darkness.' Rev. of *Licht und Dunkel in der frühgriechischen Dichtung*, by Dieter Bremer." *The Classical Review* (New Series) 29.1 (1979): 1–3.
Deacon, Terrence. "The Aesthetic Faculty." In *The Artful Mind: Cognitive Science and the Riddle of Human Creativity*, ed. Mark Turner. New York: Oxford University Press, 2006, 21–53.
DeLacy, Phillip. "Distant Views: The Imagery of Lucretius 2." In *Lucretius*, ed. Monica R. Gale. New York: Oxford University Press, 2007.
Derrida, Jacques. "White Mythology." *New Literary History* 6.1 (1974): 5–74.
A Dictionary of Greek and Roman Geography. Edited by William Smith. 2 vols. New York: AMS Press, 1966.
Dillon, John M. *The Middle Platonists, 80 B.C. to A.D. 220*. Ithaca, NY: Cornell University Press, 1996.
Dodd, C. H. *The Interpretation of the Fourth Gospel*. Cambridge: Cambridge University Press, 1965.

Donald, Merlin. "Art and Cognitive Evolution." In *The Artful Mind: Cognitive Science and the Riddle of Human Creativity*, ed. Mark Turner. New York: Oxford University Press, 2006, 3–20.

Dorigo, Wladimiro. *Late Roman Painting*. Translated by James Cleugh and John Warrington. New York: Praeger, 1971.

Drake, H. A. "The Impact of Constantine on Christianity." In *The Cambridge Companion to the Age of Constantine*, ed. Noel Lenski. Cambridge: Cambridge University Press, 2006.

Duhaime, Jean. "Light and Darkness." In *Encyclopedia of the Dead Sea Scrolls*, ed. Lawrence H. Schiffman and James Vanderkam. Oxford: Oxford University Press, 2008. E-reference edition, Monmouth University, September 18, 2012.

Dunbabin, Katherine M. D. *Mosaics of the Greek and Roman World*. Cambridge: Cambridge University Press, 1999.

Eliade, Mircea. "Spirit, Light, and Seed." *History of Religions* 11.1 (1971): 1–30.

———. *The Two and the One*. Translated by J. M. Cohen. Chicago: University of Chicago Press, 1962.

Elm, Susanna. "Julian the Writer and His Audience." In *Emperor and Author: The Writings of Julian "the Apostate,"* ed. Nicholas J. Brian-Baker and Shaun Tougher. Swansea: Classical Press of Wales, 2012.

Elsner, Jaś. *Art and the Roman Viewer: The Transformation of Art from the Pagan World to Christianity*. Cambridge: Cambridge University Press, 1995.

Esolen, Anthony M. Introduction to *Lucretius: On the Nature of Things*. Baltimore, MD: Johns Hopkins University Press, 1995.

Everitt, Anthony. *Cicero: The Life and Times*. New York: Random House, 2003.

Filoramo, Giovanni. *A History of Gnosticism*. Translated by Anthony Alcock. Cambridge, MA: Basil Blackwell, 1990.

Finamore, John F. "Iamblichus on Light and the Transparent." In *The Divine Iamblichus: Philosopher and Man of Gods*, ed. H. J. Blumenthal and E. G. Clark. London: Bristol Classical P. 1993.

Francis, James A. "Visual and Verbal Representation: Image, Text, Person, and Power." In *A Companion to Late Antiquity*, ed. Philip Rousseau. Malden, MA: Wiley–Blackwell, 2009, 285–305.

Gale, Monica R., ed. *Lucretius*. New York: Oxford University Press, 2007.

Gerson, Lloyd P, ed. *The Cambridge Companion to Plotinus*. Cambridge: Cambridge University Press, 1996.

Gillespie, Stuart, and Philip Hardie, eds. *The Cambridge Companion to Lucretius*. Cambridge: Cambridge University Press, 2007.

Goldhill, Simon. "Refracting Classical Vision: Changing Cultures of Viewing." In *Vision in Context*, ed. Teresa Brennan and Martin Jay. New York: Routledge, 1996.

Gombrich, E. H. "The Heritage of Apelles." In *The Heritage of Apelles: Studies in the Art of the Renaissance*. Ithaca, NY: Cornell University Press, 1976.

Goodenough, Erwin R. *By Light, Light: The Mystic Gospel of Hellenistic Judaism*. New Haven: Yale University Press, 1935.

The Gospel According to John. Translated by Raymond E. Brown. The Anchor Bible, vols. 29 & 29a. New York: Doubleday, 1966.

The Gospel of Thomas. Edited and translated by Marin Meyer. San Francisco: Harper & Row, 1992.

Grandy, David. "The Otherness of Light: Einstein and Levinas." *Postmodern Culture* 12.1 (September 2001): sections 1–47. Accessed August 17, 2011. http://pmc.iath.virginia.edu/issue.901/12.1grandy.html.

Hanfmann, George M. A. "Hellenistic Art." In *Readings in Art History*, ed. Harold Spencer. New York: Charles Scribner's Sons, 1982. I: 129–47.

Harley-McGowan, Felicity. "Christianity and the Transformation of Classical Art." In *A Companion to Late Antiquity*. Malden, MA: Wiley–Blackwell, 2009, 306–26.

Henig, Martin. *A Handbook of Roman Art*. Ithaca, NY: Cornell University Press, 1983.

Herman, Arthur. *The Cave and the Light*. New York: Random House, 2013.

Hijmans, Steven Ernst. *Sol: The Sun in the Art and Religions of Rome*. 2 vols. Groninger, Netherlands: University of Groningen, 2009.

Hillar, Marian. "Philo of Alexandria." *Internet Encyclopedia of Philosophy*. Accessed October 16, 2013. http://www.iep.utm.edu/philo/.

Horace. *The Odes*. Translated by David Ferry. New York: Farrar, Straus, and Giroux, 1997.

———. *The Epistles of Horace*. Translated by David Ferry. New York: Farrar, Straus and Giroux, 2002.

Jensen, Robin M. *Understanding Early Christian Art*. New York: Routledge, 2000.

Jonas, Hans. *The Gnostic Religion*. Boston: Beacon, 1958.

Julian, Emperor. *The Works of the Emperor Julian*. Vol. 1, Oration IV. Translated by Emily Wilmer Cave Wright. Cambridge, MA: Loeb Classical Library, 1913.

Kähler, Heinz. *The Art of Rome and Her Empire*. Rev. ed. Translated by J. R. Foster. New York: Greystone, 1965.

Kearney, Richard. *Poetics of Imagining: From Husserl to Lyotard.* New York: HarperCollins, 1991.

Keizer, Heleen M. Review of *Philo of Alexandria,* by P. Borgen. *Mnemosyne* (Fourth Series) 54.5 (October 2001): 593–96.

King, Karen L. *What is Gnosticism?* Cambridge: Harvard University Press, 2003.

Kleinbauer, W. Eugene. *Early Christian and Byzantine Architecture: An Annotated Bibliography and Historiography*. Boston: G.K. Hall, 1992.

Leach, Eleanor Winsor. *The Social Life of Painting in Ancient Rome and on the Bay of Naples*. Cambridge: Cambridge University Press, 2004.

Lenski, Noel. "The Reign of Constantine." In *The Cambridge Companion to the Age of Constantine*, ed. Noel Lenski. Cambridge: Cambridge University Press, 2006.

Lim, Richard. "Christianization, Secularization, Public Life." In *A Companion to Late Antiquity*, ed. Philip Rousseau. Malden, MA: Wiley–Blackwell, 2009.

Ling, Roger. *Roman Painting*. New York: Cambridge University Press, 1991.

Lucae, Richard. "On the Meaning and Power of Space in Architecture." In *An Anthology From Vitruvius to 1870*, Vol 1, ed. Harry Francis Mallgrave. Malden, MA: Blackwell, 2006.

Lucretius. *Lucretius: On the Nature of Things*. Edited and translated by Anthony M. Esolen. Baltimore, MD: Johns Hopkins University Press, 1995.

MacDonald, William Lloyd. *The Architecture of the Roman Empire*. 2 vols. New Haven, CT: Yale University Press, 1982.

———. *The Pantheon: Design, Meaning, and Progeny*. Second ed. Cambridge, MA: Harvard University Press, 2002.

Maiuri, Amedeo. *Roman Painting*. Translated by Stuart Gilbert. Geneva: Albert Skira, 1953.

Markus, R. A. "Augustine." *The Cambridge History of Later Greek and Early Medieval Philosophy*. 1967.

Mazzoleni, Donatella. "Architecture and Illusionistic Painting in the Roman House." In *Domus: Wall Painting in the Roman House*, ed. Donatella Mazzoleni. Los Angeles: The J. Paul Getty Museum, 2004, 7–39.

Meiss, Millard. *The Great Age of Fresco*. New York: George Braziller (in association with the Metropolitan Museum of Art), 1970.

Miles, Margaret. *Plotinus: On Body and Beauty*. Malden, MA: Blackwell, 1999.

Minadeo, Richard. *The Lyre of Science; Form and Meaning in Lucretius' De rerum natura*. Detroit, MI: Wayne State University Press, 1969.

Moorman, Eric M. *Divine Interiors: Mural Paintings in Greek and Roman Sanctuaries*. Amsterdam: Amsterdam University Press, 2011.

Murdoch, Adrian. *The Last Pagan: Julian the Apostate and the Death of the Ancient World.* New York: Sutton, 2003.

Needham, Joseph. *Science and Civilization in China*. 5 vols. Cambridge: Cambridge University Press, 1988.

O'Meara, Dominic J. *Plotinus: An Introduction to the Enneads*. New York: Oxford University Press, 1993.

Ovid. *The Art of Love*. Translated by Rolfe Humphries. Bloomington: Indiana University Press, 1963.

Pagels, Elaine. *Beyond Belief: The Secret Gospel of Thomas*. New York: Random House, 2003.

Pappalardo, Umberto. *The Splendor of Roman Wall Painting*. Los Angeles: J. Paul Getty Museum, 2009.

Park, David. *The Fire within the Eye*. Princeton, NJ: Princeton University Press, 1997.

Pelikan, Jaroslav. *The Light of the World: A Basic Image in Early Christian Thought*. New York: Harper, 1962.

———. *What Has Athens to Do with Jerusalem: Timaeus and Genesis in Counterpoint*. Ann Arbor: University of Michigan Press, 1997.

Perkins, Pheme. "Gnosticism." *Encyclopedia of early Christianity*. 2 vols. Ed. Everett Ferguson, et al. New York: Garland, 1997.

Peters, W. J. T. *Landscape in Romano-Campanian Mural Painting*. Assen, Netherlands: van Gorcum, 1963.

Peterson, Norman R. *The Gospel of John and the Sociology of Light*. Valley Forge, PA: Trinity Press, 1993.

Philo of Alexandria. *The Works of Philo*. Complete and Unabridged. Translated by Charles Duke Yonge. New Updated Edition. London: Hedrickson Publishers, [1854–1855] 1995.

Plotinus. *The Enneads*. Translated by Stephen McKennna. London: Faber and Faber, 1969.

Pollitt, J. J. *The Art of Rome c. 753 B.C.-337 A.D.: Sources and Documents*. Englewood Cliffs, NJ: Prentice–Hall, 1966.

Pulver, Max. "The Experience of Light in the Gospel of John, in the 'Corpus Hermeticum,' in Gnosticism, and in the Eastern Church." In *Papers from the Eranos Yearbooks*, vol 4, ed. Joseph Campbell. Princeton, NJ: Pantheon, 1960.

Ramage, Nancy H. and Andrew. *Roman Art, Romulus to Constantine*. Englewood Cliffs, NJ: Prentice-Hall, 1991.

Rexroth, Kenneth. *Classics Revisited*. New York: New Directions, 1986.

Rorty, Richard. "Philosophy as Science, As Metaphor, and As Politics." In *The Rorty Reader*, ed. Christopher J. Voparil and Richard J. Bernstein. Malden, MA: Blackwell, 2010.

Sandmel, Samuel. *Philo of Alexandria: An Introduction*. New York: Oxford University Press, 1979.

Santayana, George. *Three Philosophical Poets: Lucretius, Dante, and Goethe*. Cambridge: Harvard University Press, 1910.

Saunders, Jason L., ed. *Greek and Roman Philosophy After Aristotle*. New York: Free Press, 1966.

Schroeder, Frederic M. *Form and Transformation: A Study in the Philosophy of Plotinus*. Montreal: McGill-Queens University Press, 1992.

———. "Light and the Active Intellect in Alexander and Plotinus." *Hermes* 112 Bd., H. 2 (Second Quarter, 1984): 239–48.

Sedley, David. "Lucretius." *Stanford Encyclopedia of Philosophy*. Fall 2013 ed., ed. Edward N. Zalta. Accessed November 13, 2013. http://plato.stanford.edu/archives/fall2013/entries/lucretius/.

Serres, Michel. *The Birth of Physics*. Translated by J. Hawkes. Edited by D. Webb. Manchester: Clinamen, 2000.

Sheldon-Williams, I. P. "The Greek Christian Platonist Tradition from the Cappadocians to Maximus and Eringena." In *The Cambridge History of Later Greek and Early Medieval Philosophy*, ed. A. H. Armstrong. Cambridge: Cambridge University Press, 1967.

Siegert, Foker. "Philo and the New Testament." In *The Cambridge Companion to Philo*, ed. Adam Kamesar. Cambridge: Cambridge University Press, 2009.

Silvestri, Domenico. "From the Eloquence of Light to the Splendor of the Word." *Semiotica* 136.1/4 (2001): 117–32.

Sinnige, Th. G. "Gnostic Influences in the Early Works of Plotinus and Augustine." In *Plotinus Amid Gnostics and Christians*, ed. David T. Runia. Amsterdam: Free University Press, 1984.

Smith, Andrew. "Julian's Hymn to King Helios: The Economical Use of Complex Neoplatonic Concepts." In *Emperor and Author: The Writings of Julian the Apostate*, ed. Nicholas J. Baker-Brian and Shaun Tougher. Swansea: Classical Press of Wales, 2012.

———. *The Gnostics*. London: Watkins, 2008.

Smith, M. F. *Lucretius, "De Rerum Natura."* Translated by W. H. D. Rouse, rev. M. F. Smith with introduction and notes. Cambridge: Harvard University Press, 1975.

Smith, Rowland B. E. *Julian's Gods: Religion and Philosophy in the Thought and Action of Julian the Apostate*. London: Routledge, 1995.

Sterling, Gregory E. "'Day One': Platonizing Exegetical Traditions of Genesis 1:1-5 in John and Jewish Authors." *The Studia Philonica Annual* 17 (2005): 118–40.

Stewart, Andrew F. *Classical Greece and the Birth of Western Art*. Cambridge: Cambridge University Press, 2008.

Stump, Eleonore, and Norman Kretzman, eds. *The Cambridge Companion to Augustine*. Cambridge: Cambridge University Press, 2001.

Summers, David. *Vision, Reflection, and Desire in Western Painting*. Chapel Hill: University of North Carolina Press, 2007.

Taylor, Rabun. *Roman Builders: A Study in Architectural Process*. Cambridge: Cambridge University Press, 2003.

Thébert, Yvon. "Private Life and domestic Architecture in Roman Africa." Translated by Arthur Goldhammer. In *A History of Private Life: I. From Pagan Rome to Byzantium*, ed. Paul Veyne. Cambridge: Harvard University Press, 1987.

Theological Dictionary of the New Testament. Edited by Gerhard Kittel and Gerhard Friedrich. Translated by Geoffrey W. Bromiley. Grand Rapids, MI: Eerdmans, 1985.

Vermeule, Cornelius. *Roman Art: Early Republic to Late Empire*. Boston: Museum of Fine Arts, 1979.

Villard, François. "Landscape, Nature and Realism." Translated by Peter Green. In *Hellenistic Art (330-50 B.C.)*, ed. Jean Charbonneaux, et al. New York: George Braziller, 1973.

———. "Light and Colour (280-150)." Translated by Peter Green. In *Hellenistic Art (330-50 B.C.)*, ed. Jean Charbonneaux, et al. New York: George Braziller, 1973.

Ward-Perkins, John B. *Roman Architecture*. New York: Harry N. Abrams, 1974.

West, David Alexander. *The Imagery and Poetry of Lucretius*. Edinburgh: Edinburgh University Press, 1969.

Wheeler, Mortimer. *Roman Art and Architecture*. New York: Praeger, 1964.

Yiengpruksawan, Mimi a Hall. "The Eyes of Michinaga in the Light of Pure Land Buddhism." In *The Presence of Light: Divine Radiance and Religious Experience*, ed. Matthew T. Kapstein. Chicago: University of Chicago Press, 2004, 227–61.

Zajonc, Arthur. *Catching the Light: The Entwined History of Light and Mind*. New York: Bantam, 1993.

Zanker, Paul. *Roman Art*. Los Angeles: J. Paul Getty Museum, 2010.

Index

About the Author

David S. Herrstrom is an independent scholar who has taught at Queens College, New York University, and most recently at Monmouth University (West Long Branch, New Jersey). His previous books include a literary study of the Gospel of John and a volume of poems. He holds a doctorate in English literature from New York University.